Foreword

A friend of mine once challenged me to sit down for a year and keep a journal of my thoughts on Kansas weather. I wasn't sure where that would lead, but thought it was a good idea, so in the winter of 2011 I put my first thoughts down and by the fall of 2012, I had acquired a number of stories that I have collected here.

I don't pretend to be a writer by trade. This book is raw, mostly self-edited and represents what my mindset was at the time of writing.

I suppose one could consider this my memoirs of being a meteorologist, for little did I know at the time I was writing this that my career as a TV weatherman was quickly coming to a close.

I write a lot about my family and friends, who undoubtedly have completely different versions of some of the same stories. While some historical research has been done on certain events, many are purely from memory and may contain inaccuracies.

The book is grouped by seasons, beginning with winter. I hope you enjoy reading it.

Mark

WINTER

A Short Weather Memory

Kansans and humans in general have a very short memory when it comes to the weather. We forget from one season to another just how cold *cold* can be. Or how hot it was just a few short months ago.

This week, we had our first true taste of Arctic air in the Sunflower State and even I, as a meteorologist, was shocked by how cold the *cold* really was. It surrounded you, penetrated you, hurt your toes and fingers and I remembered from winters past that miserable feeling that a person has when you get a chill that you just cannot get rid of without soaking in a hot bathtub.

What amazes me almost as much is how *relative* things like cold are. We topped out in the middle 30s for two days in a row with no sunshine in late November and that had us dashing from our cars to our houses or work just to stay warm. In a couple of months, a stretch of two days above freezing will actually feel *warm*. We might even forgo the coat when that happens. It is hard to imagine. Just as that 70 degree day in February feels *so* warm, a 70 degree day in July has us scrambling for jackets and blankets.

That is one thing I like about Kansas weather. The four seasons are *so* distinct that it seldom gives us a chance to get bored.

Arctic Airmass Forecasting Challenges

Dense, cold, Arctic air is a tough challenge for forecasters across the country and Kansas is no exception. Several rules of thumb about this kind of air: it is almost always colder than you or any computer model thinks it is going to be; it almost always moves faster than you or any computer model thinks (we have a saying in our weather shop, "Cold air waits for no man!"); and it almost always takes longer to "scour out" of the atmosphere than you or any computer model thinks.

In Kansas, truly cold air behaves in several strange ways. First, it has less trouble penetrating the thinner air at higher altitudes in western Kansas than in the east, so although the visual of "molasses spreading across a tabletop" is somewhat accurate, it moves much faster on the *high* side of the table than on the low side. The term we have coined for this kind of cold air is a "High Plains Plunger (HPP)." The definition that I have come up with for a *true* HPP is that the cold front must pass through Amarillo, TX before it moves through Wichita, KS. This happens time and time again. We often joke that the front got caught up in a barbed wire fence somewhere along K-96 between Wichita and Hutchinson because these kinds of fronts often go through Hutchinson *hours* before they pass through Wichita. Then it may take several *more* hours before it finally pushes through the lower, more dense air of Emporia, Topeka and eventually Chanute. Pittsburg may literally wait until the next day.

These kinds of air masses are fun to watch with their often 30 degree difference across just a few miles, but they are frustrating to forecast because just a few miles makes the difference between a 60 degree high and a 30 degree high.

Snow: The Single Hardest Thing to Forecast in Kansas

People often think that a Kansas meteorologist loses the most sleep, pulls out the most hair, and churns the most stomach acid over forecasting a tornado outbreak in the late spring. People are often wrong. The hardest thing, by an order of magnitude or two, is forecasting a winter storm.

There are several reasons for this. First and probably most importantly, expectations from the public is unreasonably high. They want to know *exactly* how much snow is going to be in their yard by 4 o'clock tomorrow afternoon. This same expectation does *not* exist for rain, thunderstorms or any other form of precipitation, but for some reason, as soon as the water in the atmosphere turns white and fluffy, everyone expects that a meteorologist should be able to tell hours or even days in advance exactly what will be squeezed out on a single point in the state. Never mind that the ratio of snow to liquid varies not only wildly in a winter but wildly even within the same storm. The "rule of thumb" is that 10 inches of snow will melt down into 1 inch of water is just an estimate. Actual conversion rates vary from 2:1 to 40:1--perhaps even more.

Now, throw on top of that the fact that snow can vary as greatly over short distances as rain can, then multiply that by ten and you can see how the problems really start to add up. Oh yeah, and did I mention that Kansas is more often than not the "battle zone" for the changeover from rain to ice to snow?

I remember a winter storm not too many years ago where I forecast 3-6" of snow for Wichita. The changeover line set up *right* across the city and in a matter of miles, northwest Wichita got 10" of snow, downtown got ice and southeast Wichita got only rain. I figure for a stretch through near-northwest Wichita about 5 blocks wide, the people thought I was a genius, while the other 300,000 residents were pretty sure I was an idiot.

And that is why, during a winter storm, the last thought on a meteorologist's mind before going to sleep is, "will snow be falling by the time I wake up?" and his very first move in the morning is to jump up, go to the window, hesitate for a second and whisper a silent prayer, then peek through the blinds to see if the ground is brown or white.

When Does a Season Begin?

Ask an astronomer or look at any calendar and you will get the same answer to this question: either the 19th, 20th or 21st of December, March, June and September, depending on when the sun either crosses over the Equator or one of the Tropics: Cancer or Capricorn.

Ask a meteorologist, and you get a different answer. Meteorological seasons begin on the first of those four months for two basic reasons. The first is ease of "bookkeeping." It is easier when comparing data to use month-to-month rather that season-to-season because the start dates of seasons vary, so you wouldn't exactly be comparing "apples-to-apples." Second is that the dates seem to correspond a little closer to the actual weather that is going on. As we know in Kansas, it gets cold and "feels" like winter *way* before a couple of days before Christmas just as the summer heat almost never waits until late June to rear its head.

This being a book about Kansas weather, the *true weather* of the seasons might follow a completely different calendar. I am not proposing we ever adopt this, but if we *really* wanted the name of the season to follow the weather of the season, I would propose that we have two long seasons and two shorter seasons. Winter and summer would both be four months long, and spring and fall, unfortunately, would only be two months long.

Although it wouldn't work out every year, I would start winter on November 20th. This would put the Thanksgiving holiday in winter where it belongs. In Kansas, by Thanksgiving, the leaves are off of the trees, there have been enough hard freezes to put most plant life to sleep for the winter. It would just describe the season better.

Spring would then start on March 20th, right about where the astronomers say. Although March seems like it *should* be one of the warmer, spring months, it often brings our biggest blizzards and snowstorms and you don't really *feel* warm and the trees don't really start to bud out until late in the month.

Summer starts May 20th, about the time kids get out of school because, let's face it, that is when summer begins in the minds of most anyway. Plus, that last week of May can get pretty warm with its tornadoes and severe weather reaching a peak; the combines come out of the shed and get fixed and greased up for the coming wheat harvest. It just *feels* like summer.

Finally, fall would begin September 20th as it always seems like the relentless August heat just holds and holds until about State Fair time when we finally get a good, defining cold front through the region, breaking summer's back for good. It is the time to put the winter wheat in the ground for next summer's harvest and it often seems that someone just "flips a switch" and we have two glorious, drawn out months ahead of comfortable, breathable, beautiful, fall weather to bring arguably the best season to Kansas.

The Relentless Kansas Wind

Many of the Plains States claim to be known for their wind, but when it comes to movement of air, Kansas takes the purple ribbon. Our state is even *named* after the Native American tribe, Kansa, who's name roughly translates to "People of the South Wind." Depending on how you measure it and which list you look at, Kansas has two of the top three windiest cities in the entire country. Topping most lists is Dodge City and Wichita usually comes in third. But why?

The first and most important reason is the lack of what meteorologists call "surface friction." In other words, we are very flat and we don't have any big forests. It surprises many that these two things can be so effective at slowing down the speed of the wind near the surface in what meteorologists call the "boundary layer," but wind, like everything else in physics, is subject to the force of friction which transfers energy and slows things down. But you might say, "This lack of surface friction could be used to describe *many* states from Texas all of the way up to the Dakotas," and this is true, but Kansas has another interesting property geographically that boosts us up just a notch over any of the other, "flat, treeless" states.

First of all, wind is just the atmosphere's way of trying to equal out pressure by rushing air from an area of high pressure to an area of low pressure. The more intense these two are and the closer together they are, all other things being equal, the higher the wind. Meteorologists call this a "pressure gradient." Kansas and much of the Plains often sit to the west of what is called the

"Bermuda High," an area of high pressure which frequently migrates to the southeastern United States and just hangs out there for months. If you have ever been to Atlanta, you may have experienced this in their stifling heat and humidity with relatively light winds. To the west of Kansas is the Rocky Mountains, and due to something you learned in seventh grade science called the "Bernoulli Principle," whenever the jet stream blows over the Rocky Mountains, you get an area of low pressure to form in their "Lee" or down-wind side--the same thing that causes "lift" on an airplane wing. What makes Kansas unique is what makes the drive across eastern Colorado so spectacular: you drive for miles and miles on relatively flat, treeless ground until in the distance you see giant mountains rising out of the plains. Nowhere else along the Rocky Mountain range is the changeover so dramatic. Think about Colorado Springs or Denver where the eastern part of the city looks a lot like Kansas but the western part of the city sits firmly in the mountains. This quick change-over from plains to mountains allows that lee-side low pressure to develop very rapidly and "bomb" very deeply which increases the pressure gradient bringing higher wind. Looking at a weather map of the United States, you will almost always see this area of low pressure, or an elongated area of low pressure abbreviated as a "trof" and depicted by a dashed line, somewhere from northeastern New Mexico to southeastern Wyoming. In other words, Kansas is precisely and uniquely placed to receive more wind than any other state.

There Are Few Certainties in Weather

A wise meteorology teacher of mine at the University of Kansas advised us in one of our forecasting classes, "Never forecast a 100% or 0% chance of anything in Kansas!" His point being that even if something looks like an absolute certainty, the atmosphere is still too unpredictable to go that far. Just a few days ago, the forecast from the National Weather Service called for a 100% chance of rain 4 days out. They must have had a different teacher than I did.

Statistical Probabilities in Weather

We have heard it all of our lives growing up in Kansas, "20% chance of rain," but what does that really mean? Social scientists have found time and time again that the majority of people cannot grasp statistical probabilities. This is how the lottery stays in business. As it turns out, many of the very people that *use* statistical probabilities also have no idea what it means. This is especially true in weather. A "percent chance of rain" is arrived at by taking the anticipated areal coverage for a forecast area and multiplying it by the forecaster's confidence. Areal coverage refers to the amount of an area that will receive measurable precipitation. For example, if a forecast area is 1000 square miles (the "official" definition of a county in Kansas is 1296 square miles, and that is a typical forecast area) and 400 of those square miles receives moisture, that would be a 40% areal coverage. Thus, a 20% chance can mean an almost infinite combination of things:

Areal Coverage	Forecaster Confidence	Probability of Precipitation
100%	20%	20%
20%	100%	20%
80%	25%	20%
25%	80%	20%
50%	40%	20%
40%	50%	20%

No wonder there is so much confusion over what it means. There was actually a heated debate in the early years of forecasting whether probabilities should even be included at all. In recent years, the pendulum has swung the other way, and now probabilities are being proposed to be used in everything, including tornado warnings. A warning in the near future may actually say, "There is a 2% chance of a tornado striking your house." Statistically, this is a *huge* number, but as I mentioned before, people don't understand statistics and might take this as "too small of a chance to worry about."

For many forecasters, the probability of precipitation has become a way of "hedging your bet" instead of being used in the scientific way it was proposed. This is why you hear "20% chance" in your forecasts so often. What makes this all even more absurd is that the basic question of "is it going to rain at my house?" is really a binary one: the answer, in the end, is either yes or no; 100% or 0%. So the next time you listen to a forecast and try to "decipher the code" of the forecaster in question, listen for some key words: chance, scattered, likely, widespread, locally heavy, etc. These words are in plain English, the way Kansas people talk, and probably give you a better idea of what the weather is really going to be like.

On a Clear Day, You Can See Cleveland

Part of my love of weather and the Kansas sky came from growing up on a farm where trees didn't block my view of the horizon. We Kansans like to see our storms coming from a *long* ways away.

A little over two miles from where I grew up, there is an unassuming geographical feature we simply called, "The Hill." It is a gentle rise in southwestern Kingman County on the south side of the Ninnescah river valley along the Zenda blacktop with no trees and from which you can see for twenty miles or more across three Kansas counties. It is quite a vista for a young weather-enthusiast's mind and when we were children we would go up to "The Hill" on the 4th of July and watch the fireworks shows from Kingman, Cunningham, Pratt and several other small towns (or farms with a larger fireworks budget than ours). On a clear day, my father used to like to go up there and point out all of the grain elevators he could see and try and name them, "Cleveland, Penalosa, Rago, Zenda, Cunningham, Kingman, Pratt, Preston, Belmont..." And there were always a couple that we couldn't quite figure out or would argue about way off on the northern horizon.

Now that I am a grown, professional meteorologist with children of my own, I try to take them there occasionally to watch a storm drift across the valley or to see a meteor shower, comet or lunar eclipse. You can probably see fifty times the numbers of stars here that you can see in Wichita since it is far enough from the city's "light pollution." I have dreamt that if I ever won the lottery, I would buy that piece of ground and build a weather/sky observatory there, but I know that would just ruin some of the mystique of an unassuming rise on the Kansas plains; a spot from where humans have likely been stopping for thousands of years to watch a storm, spot a herd of buffalo or simply admire the view.

When Warm Air Does Not Mean Warmer Temperatures

At least once a year and usually more than once, every forecaster in Kansas gets burned by something called "Warm Air Advection" (WAA). It is simply a term used to describe warmer air moving into a region replacing relatively colder air. The problem with this is that it seems pretty straightforward: warmer air is moving in, so the high temperatures will be going up. Unfortunately, Kansas weather is very seldom straightforward. While WAA *can* mean warmer temperatures, and often does, it can also mean another thing: unrelenting stratus clouds. Stratus clouds are the "blanket" clouds that are low and gray and cover the entire sky with no hint of sunshine breaking through. What happens with WAA sometimes is that what looks like warmer temperatures moving in to bring a sunny, warm day is actually the atmosphere's way of disguising the fact that warm air *rises* and is sometimes *lifted* over a shallow layer of cold air. This often happens in the late fall and the early spring and as long as I have been forecasting, I have been burnt by this several times every year. This does not make the people who trust my forecast very happy. The forecast goes something like this, "It looks like a great weekend ahead with plentiful sunshine and after our highs in the 30s today, we will climb to the 40s and even a few 50s this weekend...a great time to get out and enjoy!" Then the weekend goes something like this: a brief hour of sunshine early Saturday morning with clear skies allowing the temperature to drop as far as possible, then low clouds appear on the southern horizon seemingly out of nowhere, move across the sky, block out the sun, and the temperature stops somewhere in the lower 30s and hangs out there all day; and the next day; and the next.

It is a miserable feeling for a forecaster to be *so* wrong and what adds insult to injury is that the stratus deck is often narrow or small, and one county away, they *are* sunny and 15 degrees warmer. To those viewers, you are a hero; to the 500,000 viewers under the deck of clouds that were planning on spending some time on their *own* deck, your name is mud.

There are a few tools and some research that has come out in recent years to help prevent this sort of forecast "bust," but like everything in Kansas weather, they only work sometimes, and just when you think you have it figured out, the exact opposite happens, making you look foolish again.

So the next time this happens to your local, Kansas forecaster, try to cut him or her a little slack and realize that it simply happens to *every* Kansas forecaster and has been going on as long as there has *been* a weather forecaster in Kansas and our children's children's forecasters will *still* be getting burnt by the "warm air that isn't."

Christmas Eve, 1983

The coldest Christmas Eve in Kansas history was one that will be forever burned in my mind as the closest I came to death from the weather. At the National Weather Service in Wichita, a record was set for the most consecutive days with low temperatures below zero, every morning from the 18th through the 25th. The month ended with ten days of sub-zero lows, a record that stands to this day. The average temperature that month was a mere 16.3 degrees, undercutting the next closest record by an astounding 7.6 degrees.

The Christmas Eve tradition in our family went something like this: the day was spent getting ready for Santa to come and pay our farmhouse a visit, and that meant a lot of things needed to be done. We kids awoke earlier than usual and mom had a list of chores for the day: clean the house, clean the fireplace, light a fire for the evening celebrations, find a *clean* sock with *no holes* in it to hang on the mantle over the fireplace (the bigger the better to hold more chocolates, nuts, candies and fruits), put out the chairs and put our names on them so Santa would know which chair to put our presents on, pack up homemade Christmas cookies for all of the neighbors, help make the traditional Christmas Eve supper of Irish potato soup and do it all while on our best behavior.

In our house, Santa came always on Christmas Eve before Midnight Mass while we were out delivering cookies to all of the neighbors, and he never wrapped the presents, he just placed them on our chairs, arranged from oldest to youngest.

Christmas Eve morning on the farm in 1983 dawned at thirteen degrees below zero Fahrenheit with a howling north wind and a few inches of snow on the ground, but the snow had fallen several days before and crusted over, so most of the blowing and drifting was done. The National Weather Service in Wichita recorded the highest barometric pressure ever on that Christmas Eve day at 31.21" (1057.8 millibars); something that can only come from incredibly dense, cold air.

There was debate as the afternoon drew on as to whether or not it was even safe for us to go deliver cookies to the neighbors, but my brother, 22 at the time assured mom and dad that they could get through the cold and that the roads had all been cleared by now so there was nothing to worry about. We kids (all except my oldest sister, 20, who stayed behind to listen for Santa with mom and dad) loaded up in the old family sedan with blankets and warm hearts and set out to make our annual delivery to the neighbors while singing Christmas carols and looking out the window for any sign of Rudolph's red nose (you could tell the "real" one from an airplane because it didn't blink). We decided the best route to go would be to hit the McAdams first (the Wood's weren't home) before swinging back over to the highway to hit the Harpers and Giefers and hopefully that would give Santa enough time to get down the chimney and back out without being seen. We made it to Goff's corner just fine and then turned west to take the road to the Zenda blacktop with the north wind at our side. As we approached the hedge row close to the highway, my brother came to a stop. It looked in the headlights like the road might have drifted back shut during the day after all. "We can make it!" my brother decided and he gunned it, plowing straight into snowdrifts that quickly engulfed the car and brought it to a halt. We were stuck a little over a mile from home and almost two miles short of the McAdams. I don't remember feeling panic even though I probably should have given the fact that people die like this every winter. I guess those thoughts don't occur to a 16 year-old boy. It was decided right away that my brother and I would walk back to the house, get dad's pickup and come back and rescue the sisters. My youngest sister was eight at the time and it was she that we were worried about the most. As my brother and I stepped out into the drifts, the biting north wind took our breaths away. We only hesitated a moment before deciding that it would be quicker to go directly across the wheat fields to the house (although if we wouldn't have made it, they may not have found us until Christmas afternoon) and that running would be smarter than walking (even though it meant taking deeper breaths of the icy air). Fortunately, the sky had cleared and the wind had died down enough that we could see the farm's lights in the distance. Unfortunately, since the sky had cleared, the temperature had dropped even farther and with the remaining wind chill, we were in danger. We ran for what seemed like an hour, our lungs burning in the ice-cold air, until we reached the farm, ran inside and told our parents what had happened. We started the pickup and drove the long way around two sections to get back to where we had left the girls and by the time we got there, some were crying and frostbite was beginning to set in. We walked through about one hundred yards of snowdrifts to get them, then walked with them back to the pickup, and somehow, the six of us fit in the front seat and we made our way back home.

The cookies didn't get delivered that year, but somehow God was watching over us and we all survived that Christmas relatively unscathed, although some of my sisters still have the pain of early frostbite whenever their toes get cold even to this day.

Now that I live in the city, it is easy to forget how vulnerable you can be growing up on a farm where the houses are far apart, the roads aren't always well maintained and something as innocent as delivering Christmas cookies to your neighbors can bring you face-to-face with your own mortality.

December, 1989

The 80s were a time of extreme temperatures in Kansas. From the extreme heat-wave of the summer of 1980 to the Christmas Arctic outbreak of 1983 to the pre-Christmas Arctic outbreak of 1989 we earned our "tough" stripes that decade!

In the fall semester of 1989, I was 22 years old and a senior at The University of Kansas at Lawrence, finishing up my Atmospheric Science degree. I was a Teaching Assistant (TA) in the Physics and Astronomy Department and was the official, paid weather observer for the University of Kansas weather station. At that time, the weather station was located on the 6th floor of Malott Hall and the weather instruments were mounted on the roof to avoid being tampered with by mischievous students.

December of that year brought the second Arctic outbreak in just six years and I remember the stinging cold of the biting north wind as I went up to the roof to collect the rain cylinder that was used to collect snow for melting down to find the moisture content. Observations were to be taken at 7AM and 7PM local time, then called in to the National Weather Service in Topeka and Bryan Busby at his TV station over in Kansas City. In order to speed up the snow-melting process, we had an old hair drier in the weather lab, but the stainless steel cylinder would get so cold that it would take close to an hour of being indoors to finally warm the snow enough to melt it all down to liquid to be poured into the internal tube for measuring. The "official" readings taken at the National Weather Service that month included an impressive -24 on the morning of the 3rd and on December 14th, the high only reached -7.

I was active in the St. Lawrence Catholic Campus Center Choir and 1989 saw the first year of what was to become a yearly tradition called "Lessons and Carols" with readings from scripture interspersed with Christmas hymns, chants and carols. A day or two before this, a KU tradition was held at the old Hoch Auditorium (the one that would later be struck by lightning and burned to the ground) called "Christmas Vespers" where the choirs would sing and a brass choir would meet those coming in the front door as it played from the balcony overhead. Despite the bone-chilling cold, the brass players somehow managed to play the tunes without their valves and slides freezing up and without their lips freezing to their mouthpieces. When "Lessons and Carols" at St. Lawrence rolled around, we were debating cancelling it because it was so cold that it was dangerous and we weren't sure anyone would brave the weather and actually show up. I remember walking up to the doors of the St. Lawrence Center in the late afternoon with the snow drifting and skifting around pretty sure that nobody would brave those roads and that biting wind, but by the time the first notes were sounded on the beautiful pipe organ, the chapel was filled with warm bodies bundled up against the cold and a tradition was born.

Finals being done, it was time to head home for the holidays. My sister was attending KU with me and given the fact that we both drove old, broken-down cars with leaking heater cores, we decided to wait until the next day to brave the drive home. Unfortunately, the skies cleared that night and by the next morning, it was dangerously cold with bitter wind chills as we drove south out of Lawrence toward Ottawa to catch I-35 to Emporia.

I was in a 1968 Ford Fairlane with no heater that had been crashed to the point that you could look out the crack where the driver-side door was supposed to meet the body and watch the road passing by. My sister was in a 1980 Dodge Aries K-Car without a heater that had once been owned, ironically enough, by Wichita broadcasting legend, Meteorologist Jim O'Donnell. We both bundled up with everything we had: multiple pairs of socks, shoes covered by rubber boots, long-johns, multiple shirts, coats, hats, gloves and scarves and wrapped ourselves in a blanket cocoon before hitting the road. We made it to one of those gas stations that you see off to the north of I-35 somewhere between Ottawa and Emporia before we couldn't take it anymore and we signaled each other to pull off and go inside to warm up. We bought a warm drink, but mostly we just stood and sat there until we could feel our fingers and toes again hoping against all hope that the temperature would miraculously warm up before we had to hit the road again. Reinvigorated by the warmth and hot chocolate, we hit the road again, but by the time we reached Peabody on US-50, we were again numb and in pain from the cold and took another long stop in the "Stop-and-Shop" there. One last leg of the bitterly cold trip finally took us home to the farm west of Kingman.

I often think of the cold of that December, that trip and how we really shouldn't have even been out on the road but when the draw of a warm, home Christmas on the farm is beckoning and finals are over, you make calls that make more sense to the heart than the head.

"How Come it *Never* Snows Like it Did When I Was a Kid?"

This is probably the second biggest psychological misconception that I get asked about repeatedly, the first being, "How come the rain *always* misses us?"

I believe there are actually several things at work in the human brain that causes one to feel this way. The first being, when you were young, you were a lot smaller, and a significant snow, like 12", seemed much deeper than a similar snow would now.

The second one is harder to get the mind around, and it may have an actual, fancy psychological term that I am unaware of, but it goes something like this: one or two events in the very impressionable times of your youth tend to get spread out into an "*always*" in the adult mind. In other words, it just *feels* like every winter when you were a kid, you got big snows that lasted for weeks and you built all sorts of snowmen and sledded and played in the snow drifts, when in all reality, if you were able to go back and look, that probably really only happened in one or two major events. It is just that they were so much fun, you want to believe that they happened all of the time.

Statistics back this up somewhat with a few caveats. If it really did snow more "when you were a kid," then the snowiest winters in history should be longer ago with a steadily decreasing amount of snow each winter since. Looking at the records, there is a small clustering of them in the 1970s, when we seemed to go through an unusually cold period, so if you were a child in the '70s, there may be at least a little truth to your perception. Conversely, in the 1990s we tended to have more of the "low snow" records in what turned out to be a fairly dry, warm decade. The problem is, I hear this same question from people that are 80 to 18 and all of them were certainly not children in the 70s and some were, in fact, children in the 90s.

Finally, snow varies greatly over just a short distance, so someone like me who grew up in Kingman may have seen several heavier snows that were only minor snows or rain in Wichita, where I live now, and vice versa.

Some snow statistics that might interest you and prove the above point via the National Weather Service in Wichita:

The five snowiest Wichita winters:

1911-12 39.7"
1987-88 39.4"
1959-60 36.0"
1974-75 34.4"
1972-73 30.0"

The five least snowiest Wichita winters:

1922-23 0.7"
1903-04 0.8"
1993-94 1.3"
1991-92 1.8"
1907-08 2.0"

The variability of snow from year-to-year can be dramatic even within a decade and from one winter to the next. For example, in the decade of the 2000s, the least snowy Wichita winter was the winter of 2001-02 with 4.8" while the snowiest winter came just a year later in 2002-03 with 25.7"!

Never Be Unprepared for the Kansas Cold

The time that I actually did the most damage to my body, physically, from the cold was not a remarkable time in the records of Kansas weather. It was not a major cold-air-outbreak or blizzard that actually got me hurt, it was an averagely-cold winter day when I was about 13 years old, which would have put it somewhere around 1980.

As a teen, I worked on my uncle's Hereford ranch in and around Kingman. It was a pretty sweet deal for me as not only did I get paid to check, feed and work cattle, but I also got a 5-speed Schwinn bike to ride around to check them at first, then later a 1965 Ford Galaxy 500 we affectionately called, "The Blue Bomb." The Blue Bomb had an AM radio in it and I attribute my love and appreciation for pop music from the 50s and 60s to that to this day.

This day was like any other on the ranch: after getting home from school, I changed my clothes and got ready to go out and check the cattle and see if they needed a new round bale put in their pen. There was a coating of ice on the ground and a light dusting of snow on top of that, so there was a chance, but I had just put out a bale the day before and we were only wintering a few cattle in that pen, so I figured I had one more day before I would need to do that.

As many impatient teens do, I tried to cut corners, and since I thought I was only going to be outdoors for a short amount of time, I didn't do the usual preparations of putting on two pairs of socks, my work boots, my coveralls and rubber boots over the top of my work boots to keep my feet warm and dry. For whatever reason, I decided that the one pair of socks and tennis shoes I had worn to school that day would suffice.

I believe it was probably about 20 degrees when I walked out the back door of the house with overcast skies and about a 20 mph wind blowing from the north. Again, nothing remarkable for a typical Kansas winter. No sooner had I walked out the back door, and my uncle/boss pulled in across the road where we kept the cattle and bales and I went to see what he was up to. Maybe he was going to do my chores for me and I would have an even shorter stay outdoors than I thought!

For whatever reason, he was in a sour mood and immediately started getting on me for not getting my chores done earlier and, "why didn't the cattle have a fresh bale already?" The antique Alice Chalmers WD-45 that we used to feed bales with was down for repairs, so my uncle decided that we needed to start our old John Deere 4010 and use it, even though it had a "summer blend" of diesel in it and would be hard to start. Sure enough, the battery was dead and after hooking it up to jump-start it, we commenced to using the better part of a can of ether to try and get it to fire.

By now, the cold was starting to set in to my under-covered body and with the mood my uncle was in, I didn't dare ask if I could go back to the house to warm up and put on some more clothes. I was starting to shiver from the cold and my extremities were hurting, especially my toes. It was then decided that we should try to "pull-start" it, a method in which you get the tractor up to some speed by pulling it with the transmission in gear, then you "pop the clutch" and transfer the momentum from the turning wheels to the engine. This works remarkably well in the right circumstances, but these were *not* the right circumstances.

With my uncle in the pulling vehicle and me on the tractor, we headed south with the wind at our back. After trying multiple times over about a quarter of a mile, we headed back north and that was the final straw in my tolerance of the cold. I was standing on a cold, steel plate covered in ice and snow with my face sticking up high into the north wind that was now stronger thanks to the added forward speed we added by pulling the tractor into it. By the time we got back to the farm, tears were streaming down my face and I knew that something was terribly wrong. My feet had gone from the cold, stinging sensations that I had felt before when they got too cold to a deep, penetrating pain that was shooting up my legs.

I had trouble stepping down off of the tractor (which never did start, by the way) and explained to my stoic, German uncle that the tears on my face were just because of the cold air blowing into my eyes. Like many of his generation, heritage and upbringing, he was not the kind of man that you showed any emotion around.

He must have known something was wrong and he told me to get in the cab of the pickup to warm up, but the hot air on my feet only made them hurt worse. What happened next is kind of a blur, but I remember stumbling to the house, having a hard time walking on my damaged feet, and as soon as I got on the back porch and in the presence of my Irish mother, whom I *could* show emotion around, I started bawling and telling her that something was wrong. The wise woman that she was, she did exactly the right thing for early frostbite as she took me to the bathroom and got luke-warm water running into the tub and had me put my feet under it. The water felt as if someone were dropping a bowling ball on my feet and I cried and wailed even more. Eventually, the feeling began to come back to my feet, which were an eerie color of white that I hadn't seen on my body before and I managed to minimize the nerve damage from what was undoubtedly early frostbite.

To this day, whenever I put my feet into warm water in a bathtub, they turn white and that tingling feeling comes back to me of a day many years ago when I went out of the house unprepared for what a "typical Kansas winter day" could throw at me.

The Thermodynamics of a Late-February Snow Drift

Thermodynamics is a branch of study of science and engineering which describes how heat relates to things like energy and work using variables such as temperature, pressure and entropy. In other words, it is what a meteorologist lives and dies by. It describes everything from the heat engine of a 1,000 cubic-mile Kansas supercell thunderstorm to how often your furnace needs to kick on during an Arctic outbreak.

One thing that fascinates me especially is the microscale thermodynamics that go on around a late-February Kansas snow drift. By late February, there are typically patches of snow across the state that seem to just refuse to melt, usually found in ditches, fence rows, tree rows, or at the edge of the Wal-Mart parking lot. Also in late February, the thermometer always hits 70 or very close to it.

When I was home from college one year with my college roommate, who was studying architectural engineering, we stood near a snow pile across the road on the farm by the barn on one such day and marveled at what we were experiencing. Here we were in short sleeves, feeling the warmth of the sun on our skin, and less than a meter away was *ice*. That means that at some point across that meter, and probably in a matter of mere centimeters or even millimeters, the temperature drop is about 50 degrees. Think about it; it is *one* impressive thing to see a 50 degree temperature drop across the 400 miles from one side of Kansas to the other, but take that same temperature drop and spread it across a distance less than the width of your *hand*!

There *is* melting going on in that drift or pile of snow, but it is shockingly slower than one would think in such conditions. Part of what is going on is that those snow crystals that fell days, weeks or months ago have essentially glaciated, becoming one solid mass of ice crystals, so that the drift, rather than acting like a bunch of individual snowflakes, actually acts like a single, solid block of ice and have created their own local climate. Think about dropping an ice cube from your refrigerator on to your warm kitchen floor. It doesn't melt immediately, but in a matter of a half an hour or so, you are stepping in a cold puddle of water in your favorite dry socks, wishing you had picked it up when it first dropped. This Kansas snow drift "block of ice" can actually survive for hours and even *days* in similar conditions. It is fascinating.

I would love to have the time and knowledge to actually dedicate to writing a master's thesis on the microclimate around a late-

February Kansas snow drift. Come to think of it, that would actually be my second choice. My first choice would be a master's thesis on an August, afternoon, Kansas "dust devil." Since I have no actual desire to ever sit in front of a thesis jury, I will just have to be content with being fascinated by some of those small, Kansas weather phenomena that capture a meteorologist's imagination.

The Two Things That Happen in Kansas Weather Every Year but Everyone Acts Surprised About Them

There are two things that have happened in Kansas weather every year since I became a meteorologist (although they have likely been going on forever). Every year it hits 70 in February and every year it snows in April. Every year people are shocked by these two events and I get many questions from both the public and the media about what was causing this "unusual" weather!

Keep in mind, I am not saying that this happens in *every* location in Kansas every year, but somewhere in the state these things happen and you can pretty much put money in the bank on it. I suppose there has been some odd year where the cold or snow were just too stubborn for even Elkhart to hit 70 or it was too warm too early or too dry for even St. Francis to get snow, but these years would certainly be the *true* rarities.

Some of the "old timers" actually relied on both of these phenomena and planned their farming and gardening practices around them. The warm day in February was used to sow the oats that would be needed for horse feed come summer. The garden plot was often worked up and peas and strawberries were planted. The old saying went that if your peas weren't snowed on at least once, you wouldn't even raise enough to make creamed peas and potatoes come May (a true Kansas delicacy...if you have never had fresh peas and baby potatoes straight from the garden in a cream sauce, I am not sure you can say that you have lived a complete life).

The April snow, especially in the northwest part of the state, is an important source of moisture (and snow for those peas) in a part of the state that can use all of the moisture it can get. Since the last freeze in the northwest doesn't occur until after mid-April, there is usually no damage from this "late season" snow, and it is welcomed!

So the next time your backyard thermometer hits 70 in February and you have your windows open, airing out the germs from the winter or the next time snow falls after you have already turned the calendar to April and you are thinking more about flip-flops than your winter coat, resist the urge to e-mail your Kansas meteorologist and ask them, "what is causing this *unusual*, Kansas weather?"

My Earliest Weather Memory

I only have two very distinct memories from before I turned five years old. One is of my Irish grandfather pulling up in his blue pickup to bring us 7-Up and M&Ms when we were sick and the other is of the February Blizzard of '71.

I was only three and a half years old at the time and the thing I remember about the blizzard was the depth of the snow. I have a very distinct picture in my mind of my older sister taking my hand and taking me outside, all bundled up, to walk in the pathway that my father had shoveled so that he could get out and do chores. I remember walking down what looked like a snow tunnel, only open on the top and the snow was too high for me to see over it. I remember someone lifting me up to look over the top and I was looking south from in front of our farmhouse and all I could see was snow, snow, snow.

Not a very impressive memory, perhaps, but even to this day, I hear people tell of the two big blizzards of the early 70s. There was another one in late February of '73 that I remember much more clearly. I was five years old then and the blizzard occurred right around my dad's birthday, February 21st. He had the flu at the time and we were snowed in for days and days. We used to spend a lot of our winter evenings in the farmhouse by the fireplace doing jigsaw puzzles while listening to John Denver, Jimmie Rogers, Anne Murray and Crystal Gayle on the mono record player. During this blizzard, we were doing an especially hard jigsaw of a picture of the "Lincoln Memorial" taken at night. Every piece was a shade of black, gray or brown and each one looked like the other. After that blizzard was over, we never did that puzzle again because it reminded my dad of how sick he was, being all cooped-in, and having to still do chores like feeding hogs and cattle throughout. My dad was a strong, independent man who almost never got sick. He would have been turning 37 that year, in the prime of his life, and his only help was my brother, who would have been 11 at the time. Most of the cattle feed that year was in old-fashioned shocks from an antique binder that my dad co-owned with a couple of the neighbors. The shocks (think little "teepees" of bundles) were left out in the field. The advantage to this method was that the grain in the heads of the feed remained intact and the stalks remained un-crushed, unlike in modern baling equipment, and storing it in shocks meant that it shed the water better and the feed stayed higher quality for longer with a higher sugar content to give the cattle much-needed calories. Of course, this was very labor-intensive way to do forage, which is the main reason why it is no longer done, requiring driving to each shock, picking up each bundle either by hand or pitchfork, throwing it on a hay trailer, then taking it to the feeding pens to throw out for the cattle. I can only imagine how miserable this was during a blizzard when you were running a high fever and fighting the flu. Normally, when dad or one of the neighbors got sick, the others kicked in to feed for them, but during a blizzard, you were pretty much on your own to make sure your livestock didn't suffer from hunger or thirst.

February blizzards can be real "doozies" here in Kansas. The Gulf of Mexico is starting to "open for business" again after our driest month of January has wrapped up. Storm systems coming out of the southern Rockies have more energy or "dynamics" to work with to pull up the moisture and warm air over the top of what is some of the coldest air in certain years. The all-time lows across parts of Kansas have been set in February, including the 22 below zero on February 12, 1889 in Wichita. With the increased dynamics also comes increased winds which can whip around the snow shutting down the state for days. To make matters worse, we almost always hit 70 in Kansas in the month of February, so the contrast from a "preview of spring" day to a blizzard just a few days later can be especially hard on the livestock, which often starts calving this time of the year, to the people who are all ready for winter to be over and spring to begin.

"Winter Storm" Means Different Things to Different Kansans

It seems to depend on which part of the state you live, or grew up in, as to what strikes fear in the hearts of a Kansan when they hear there is a significant winter storm coming. My father grew up on a farm between Dodge City and Cimarron near a small town named Howell and was a true western Kansas boy. My mom grew up on a dairy farm near Kingman and was more of an eastern Kansas girl. What struck fear in the heart of my father was the word, "blizzard." What struck fear in the heart of my mother were the words, "ice storm."

Not that western Kansas doesn't get ice storms, but they are much more rare than in the eastern half of the state due to elevation, and not that eastern Kansas doesn't get blizzards, they are just much more rare due to the lack of large amounts of snow combined with high winds; maybe a "once every 20 year" event, while they are almost a yearly occurrence in the western half of the state.

Growing up as a child of these two different "outlooks" on winter storms, we spent an awful lot of time preparing for the worst and spending time at home. My parents were their own "superintendents of schools" and they ultimately decided when they did and did not feel comfortable loading their seven kids on a school bus for an hour ride to school and an hour ride home. If things started to look bad, one of the thrills of my childhood was my dad showing up at the classroom door telling the teacher that he was taking me home, even though school was clearly in session. My other friends, especially my "townie" friends were always jealous that my father would do that, but he had survived western Kansas blizzards and mom had survived eastern

Kansas ice storms and they simply didn't take chances with either.

The two blizzard stories that I heard from my father were like the ones you might read in pioneer books about blizzards. As kids, we found them hard to believe because they were so stereotypical and out of our realm of experience, but stereotypes become that way for a reason; because they happen often enough that the story gets told again and again. The first story was of dad riding in a car headed west out of Dodge City and the visibility was so low that you couldn't see the end of the hood of the car, so his parents would have to roll down the window and stick their heads out occasionally so that they could see if they were even still on the road or not and to watch for markers so that they didn't miss the turn into their driveway. A missed turn could be the difference between life and death. The second was of the time that a blizzard was raging and my grandpa was worried about the chickens in the coop not having any water or feed, so he and my dad tied a rope around their waists and the other end to the barn, then started walking with a bucket between them, and they couldn't see each other and could only hear each other when they shouted over the wind. After walking in the general direction of the chicken coop, grandpa shouted to dad, "Something's wrong! We should have been there by now! We must have missed it!" So they worked their way back up the rope, scanning left and right for the chicken coop, but never did find it. After the blizzard, they went out and looked and realized they had walked right over the *top* of the chicken house which was now full of dead chickens. Years later, my brother had the chance to take a business trip through western Kansas during one of their blizzards. After years of scoffing at dad's stories, experiencing one firsthand made a believer out of him!

My mom's experience with central Kansas ice storms goes back to her childhood as well, but two stories of after she had children stick out in my mind. The first was a drive to the Kingman Christmas parade that was held during the daytime back then. I was playing in the high school band, and for whatever reason, they decided not to cancel the parade for weather that year and for whatever *other* reason I decided that I really needed to be there and my mom really needed to drive me. She refused to drive on the highway, so we took "the north road," a dirt road that runs parallel to Highway 54 one mile north where the traffic would be non-existent and we could get some traction on the dirt and rocks of the road. The freezing rain was coming down so hard and fast that there was no way the defroster and windshield wipers could keep up with it, so we had to stop every several miles and get out and scrape the ice off so we could see, and even the dirt and gravel was dangerously slick to stand on while scraping. The other, more precarious situation that involved an ice storm and my mother was when she was driving home from Kingman to the farm in the mid-70s while a steady freezing drizzle was falling, turning Highway 54 into a virtual skating rink. As she approached the stretch of highway we call "the Bermuda Triangle" of Kingman County, the stretch of road from just west of Blueberry Hill Road to just east of Highway 14 (a large amount of wrecks and deaths have occurred here due to three deceptively long, rolling hills that look like you have room to pass, but you do not), two semis slid into each other, one jackknifing and the other turning over on its flat nose and my mother locked up the brakes to avoid the collision, but on the sheet of ice, slid right into the mangled semis, in one of which the driver was seriously injured. She had my little sister in the front seat with her (this was a time before car seats and booster seats) and somehow they both managed to walk away unscathed, but after that, even the lightest freezing drizzle or freezing rain or even the *forecast* of it, and we weren't allowed to leave the house. This did not amuse some of us kids when we turned teenagers and wanted to go out on a Friday or Saturday night.

Having grown up closer to the eastern Kansas ice storms than the western Kansas blizzards, these are the ones I remember most myself. I have experienced 4 true "western Kansas style" blizzards in my lifetime in '71, '73, '83 and '89, but have seen too many ice storms to count. The biggest one at the farm came in the winter of '84/'85 when I was a junior in high school. I was working on my uncle's cattle ranch at the time and I remember going out to feed cattle in a thunderstorm while the temperature was in the 20s. It was the strangest thing to hear thunder and see lightning while everything got a thick layer of ice that I could almost watch thicken by the minute. I remember having trouble cutting the twine off of the round bales that I was feeding because the ice was too thick for my knife to go through, and when I went to pull the twine off, it took a whole layer of the outer skin of the bale with it as they had become one solid, frozen mass. The freezing rain continued into the night and I remember stepping out on the front porch to what sounded like gunshots as tree branches were snapping off under the weight of the ice in every direction. Soon, the ice became thick enough to uproot entire, 50-year-old cottonwood trees several feet in diameter and I remember feeling physically scared hearing those come crashing down. I went onto the back porch to talk to my friend on the phone about the amazing storm and while I was out there, a large limb from the giant American Elm tree in our back yard snapped and fell right on my father's pickup and the branches smashed through the windows of the porch, shattering glass where I was standing. I didn't get hurt, but I remember feeling like I was in a nightmare where a large, black monster was reaching its arms and fingers through the windows to come get me and I was so frozen with fear I couldn't run. All I could do was yell! We spent weeks, months and even years cleaning up branches and trees from that storm and if you went with me to the farm today I could point out at least two cottonwoods that came down during that storm that are laying there still.

Why I Am a Meteorologist Today

I guess I come by it honestly. My Grandmother McFadden (Sheahan) was the official weather observer for the City of Kingman for the U.S. Weather Bureau in the 1960s. Sadly, she died a month before I was born, so I never got to meet this wonderful lady, but my mom says I inherited my love of weather from her.

Growing up on a Kansas farm with a weather-wise father didn't hurt, either. I can remember him teaching me the "farmer's way" to read the weather: if the Killdeer are crying in the morning, expect rain soon; if you see a sundog in the evening, a storm is approaching; if snakes start climbing above ground level, it means a flood is on the way; if you want it to rain, leave the windows down in the pickup, leave all of your good tools laying out overnight and leave the can off of the exhaust pipe of the tractor! Not all of his weather wisdom came from folklore and wives-tales. Somehow, he understood the circulation around a mesocyclone and the "right hand rule" without ever taking a synoptic meteorology class in his life. We would come up from the basement after a particularly nasty storm had passed and he would say, "The wind has switched to the northeast now. That means the area of low pressure that the tornado would come from has moved off to the southeast. We are safe to come up from the basement." I don't know *how* he knew it, he just did. And I wanted to know, too. But I also wanted to know *why*.

My love for weather was solidified during a particularly nasty storm that came through the farm at harvest time between my first and second grades in school. My dad called the storm a "northerner." I know now that it was technically a "southeastward propagating, northwest flow Derecho from a Mesoscale Convective Complex" but somehow a "northerner" still sounds more "romantic" to me. This particular storm packed straight-line winds of over 100 mph that did considerable tree damage, moved around some farm equipment and blew bales of hay across the field. I witnessed that last phenomena from the front seat of the old farm pickup and it made quite an impression on me. That year, the wild rye that had come up with the wheat was quite bad, so to help control it, dad had swathed a path around the outside of every wheatfield, then baled it up into square bales and stacked them in stooks. A stooker was a machine that was pulled behind the baler that stacked nine bales, water-shedding side up in the shape of a pyramid. Being an eager, young seven-year-old, I asked my dad if I could help on the stooker, but he replied, "Oh son! One of those bales weighs as much as you do," about fifty pounds. As I watched those heavy bales of hay blow end-over-end across the field and get caught up in the fence-row, I remember saying to myself, "If *I* was out there in the field right now, *I* would be blowing end over end into the fence-row!" It is an image that has never left my mind, and from that day forward, I was fascinated by how the weather worked and why it worked that way. How could it be *so* powerful? It is just "air!"

I read every book in my school's library that I could about the weather and I am sure my teachers got tired of the papers and book reports that I wrote because they were almost always about weather! I remember asking the nun in our library one time if she had any more books about the weather (I had read them all) and she told me that I needed to try the public library in town. Kingman is blessed with a well-stocked, beautiful, brick Carnegie Library like many small towns in this part of the world, and I looked up every book on weather that I could find there. Finally, when I got to high school and it was time to start thinking about college or a career, I told my father that I wanted to be a farmer and rancher just like him. The farm economy was very bad in the mid-80s, and my wise father said, "This farm isn't big enough to support two families. Go to college and study something that you really love and when you are done, perhaps the economy will have turned around and you can start farming then." I went to my high school counselor, a wonderful man named Chris Holst, and he did some research for me and found out that the only school in the state that offered a degree in meteorology--called atmospheric science--was at the University of Kansas in Lawrence. I didn't know KU from K-State at the time and really couldn't have cared less, but I knew that if *that* was the place to go to get a degree, that is where I was going to study. I walked in the door at KU with a mission to get my Atmospheric Science degree and four years later, with no summer school (a rarity), I "walked down the hill" with a Bachelor of Arts from the College of Liberal Arts and Sciences in Atmospheric Sciences with an emphasis in Broadcasting degree and I have never looked back. I had fallen in love with the science and art of meteorology and I was destined to be a practicing meteorologist instead of a farmer.

I was fortunate to land a wonderful internship at KWCH-TV studying under Merrill Teller and John Wooldridge (he went by John Ridge on the air), then at KSNW-TV and its associated private weather company WeatherData, Inc. and studied under Mike Smith. Mike must have seen something in me that he liked because he offered me a job before I had even finished walking down the hill and I drove to Wichita on weekends to work for him.

Since TV meteorology is as much an art of storytelling as it is a science (taking complex, scientific information and paring it down to four minutes and 15 seconds of information that everyone can understand), I also had some great training in storytelling as I was growing up. Few people can tell a story like a Kansas farmer and we had some of the masters of storytelling in my neighborhood growing up. Many times I had the fortune to hear my dad and a neighbor swap stories through the window of a pickup truck, over a glass of sweet tea or standing outside church on Sunday morning. In addition to my dad, I studied "master storytelling" under Math (prounounced "Matt") Giefer, Raymond Kuszmaul, Pat McAdam, Bill Brown, Levi Rohlman and other neighbors. Those old Irish and "Deutsch" men had an uncanny ability to tell a story that would hold the interest of everyone listening--from a child to an adult. Many of the stories I heard over and over again through the years and would never tire of hearing them. If they were a funny story, even though you knew what was coming, you would laugh until your sides hurt and tears were running down your cheeks. Some of the "classics" that I can still hear those long-gone but not forgotten men tell: Pat McAdam's story about the man that painted his car with a broom; Bill Brown's story about how he was almost killed by a typhoon during World War II; Raymond Kuszmaul's story about the time they were working in the shop and the remnants of a hurricane came across Kansas; Levi Rohlman's stories about the farm equipment that he had tried that had worked and that which had failed; and the master storyteller of them all, Math Giefer. Math could tell a story with a twinkle in his eye and a wink and a nod that always kept you guessing about which part of the story was true and which was embellishment. He was a long-time, well-known auctioneer in the state and seemed to know the family history and "a story" about everyone. He was a living treasure of Kansas history. Whenever I dated this girl or that, he found out what their last name was, who their father and grandfather were and then he had a story (sometimes tragic but always with a twist of humor) about something long-forgotten.

I lament that this art of storytelling has been replaced with 140 character "tweets" and facebook status updates and YouTube videos. If I had one wish, it would be that I could tell a story one tenth as well as any of these men, including my father. The humble attempt in this book about my experiences in Kansas weather is but a glimmer of what they would have been able to tell, but I hope you, the reader, can appreciate my effort to keep the stories about Kansas weather alive.

SPRING

Priming the Pump

I grew up on a farm with a working windmill--something that is becoming a rarity these days. When I was in my teens, I helped my uncle overhaul the windmill, putting new "leathers" on the bottom of the pump rod and learned a lot about how pumping water works. The "leathers," are the parts at the bottom of the pump rod that go down into the water and basically have a one-way valve made out of leather that holds a section of water, dips down below the water table, grabs another bunch of water, and lifts it up a few feet at a time until it reaches the top to pour down the pipe into the horse tank. Occasionally, when the leathers would get old or the windmill was shut off at the wrong point in the stroke, the leathers would dry out and wouldn't seal completely enough to do their job, so "priming the pump" was necessary by grabbing a bucket of water and dumping it down into the pipe to get the leathers moistened up and give them some water pressure to get the valves to hold and water to pumping again.

March is Kansas's way of "priming the pump" for spring. The dry air of winter has essentially dried out our leathers, and we need to dump moisture back onto them in order for them to do their job creating thunderstorms that bring us the much-needed early-spring moisture. Instead of a bucket to get the moisture up from the Gulf of Mexico, nature grabs wind, a *lot* of wind in March, and cranks it up from Houston right to the Flint Hills. Although the pump carries only meager, shallow moisture at first, as it gets the ground of Texas and Oklahoma moistened up, it eventually brings deep, rich moisture with dew point temperatures in the 60s far enough north to get the severe weather machine running in high gear. By June, the pump is running full-bore and our wettest month arrives with the abundance of Hard Red Winter

Wheat and Long Stem Bluegrass that make our state the Breadbasket of the World and one of the top cattle producers in the country.

So the next time you are tempted to curse the March wind because it blows your hat off or your patio furniture over or prunes those dead branches from the trees that have been hanging around all winter, remember, without that priming of the pump, we might just turn into a desert.

Johnny

03-05-07. The day I lost my brother to suicide. It was an early March not unlike this one with some warm days and some cool ones. He was very sick at the end with his obsessions, compulsions, delusions and depression. He came over every day to sit on my front porch and smoke a cigarette and we would sometimes be cool and other times too warm, but with a west-facing porch, if the wind wasn't blowing too hard, we could soak up some sun. We talked about his meds and his illness, but we also talked about the weather, our childhood of growing up together with nine people in a three (later four) bedroom house. My entire childhood, I shared a bedroom with Johnny and he was my best friend. When I was young, we even shared a bed and I couldn't sleep without touching him (much to his annoyance). It was my way of knowing that he was still there and some monster hadn't come and kidnapped him in the middle of the night. If he was still there to protect me, I was fine and could sleep more soundly.

I miss him acutely this time of the year. It reminds me of those final days with him. The last time I saw him, on the 4th, it was one of those picture-perfect early-March days with plentiful sunshine and little wind. He had a coat on, zipped up to the top which seemed odd. I suspect he used it to conceal the .45 that he would use within hours to end the pain and suffering that he just couldn't endure for another day. The last thing I said to him as he walked toward my white, Chevy Blazer that he had borrowed because his old pickup had broken down, was, "I love you!" Thank God I had gotten into the habit of doing that to all of my loved ones years earlier, despite my stoic, German upbringing. My half-Irish blood allowed me to not be embarrassed to say, as a grown man, to another grown man those three words.

After Johnny died, I found a cluster of dried flowers in the Blazer that he took with him to his final moments. I remembered him telling me the story a couple of days before he died that his counselor had taken him for a walk along the Arkansas River on another one of those nice, early, March days and had picked them for him encouraging him to, "hang in there." While he was not able to "hang in there," he went to his death knowing that he was loved by many that he left behind.

I don't remember much about the weather the day I got the news from the Kingman County Sheriff, "We found your brother. He is dead." Nor do I remember much about the weather for the next several days or the day of his funeral and burial. It is all a haze to me, but I am sure that the weather continued to do what Kansas weather does in early March; a warm day followed by a cool one, some windy, some not, maybe a little rain. Strange thing about the weather, it goes on no matter what we are going through in life. It cares not about what we are doing or whether we are even able to pay attention to it.

Johnny chose to go "back home" to die. He left his apartment in Wichita and drove to the farm, but even in his utter despair, he was thinking of others and chose a hidden, out of the way place just over the hill out of sight and crawled under a group of Sandplum bushes on the neighbor's land, just out of ear-shot, assured that it wouldn't be a family member that would find him. One neighbor heard the shot. Another found him when he went out to feed cattle that cold, March morning.

On the two-year anniversary of Johnny's death, our family got together to go through the boxes and boxes of paperwork that we took from his apartment after his death. He was a prolific writer and we had thoughts of taking his best writing and putting it all in a book. We met at my house on what would end up being the warmest March 5th in Wichita history as the temperature climbed to 85 degrees. We had the back door open with the warm wind blowing through the screen. While we found some brilliant writing, it became obvious that his illness was even worse than most of us knew as his writings were full of rambling, angry lists and editorials. I still have the boxes of papers and perhaps some other warm, March day I will get up the courage to go through them again, whittling them down to just the best-written stuff.

March is such a time of loss and new birth, celebration and sorrow. Not only in my family where we celebrate my daughter's and baby sister's birthday along with several nieces and nephews, St. Patrick's Day (very important to us of Irish descent), the solemn season of Lent and the anniversary of my brother's death, but also in weather and nature. It is not quite warm every day nor is it quite cold every day. It rains, it snows, the wind blows, the sky is blue, the sky is gray, some flowers bloom while others refuse to, some trees green up while the wise, black walnuts know there is still an April frost coming along with a glorious Easter and the truly warm weather that lasts until fall.

I miss my brother. Not only this time of year, but always. He and I loved weather together, loved the farm together and loved growing up together. On the back of his funeral booklet, we printed the poem that he wrote for me after we passed in downtown Wichita as professional men, far from the farm where we grew up. Here is to seeing you again, my brother, in "that same yellow field" in a land where the weather is always perfect.

To My Brother

by John Bogner (1992)

Could it have been that long ago that our combines passed in that yellow field?
Eating chaff, burning oil and cutting the cheap grain, we gave each other the big "thumbs up" because life was good and the day was ours.

Today I passed you at Main and Douglas, you in your suit and I in mine.

Another "thumbs up."
Not as big or so self-assured, but a "thumbs up" just the same.

And so, my brother until I see you again with that dusty face and dark red tan, your thumbs up high in the air against the blue Kansas sky, my thumb is up to you as well.

Do not worry that we may not fit the part, for in my mind I will always see you in that same yellow field.

Early Spring "Weeds"

In a never-ending effort to homogenize their lawns into a single type of non-native plant that looks like a golf course, Kansas spend billions of dollars on herbicides, water, mowing, lawn services and seed. I can appreciate a finely trimmed, weed-free, deep-green lawn as an agricultural "work of art" as well as the next guy. A very good friend of mine and fellow meteorologist has one of those lawns that looks like something you would see in front of a fancy corporate headquarters, and as we were admiring it, I asked him, "how do you keep your lawn looking like that?" He told me, "All it takes is a lot of feed and a lot of water!" Not to mention mowing it several times a week.

Having grown up on a farm, care of the several *acres* of lawn was often an afterthought after the fieldwork and care of livestock got done. If the yard got too big to mow with the standard riding mower, it was done with a "Bush Hog" behind the small tractor. I guess it just never dawned on me to put *so* much time, effort and money into something like a lawn that I couldn't eat or sell as a commodity.

I think there is a Kansas pioneering component to it as well. My father taught me from a very young age the names of all of the flora and fauna that was found on our farm, and he usually had a story to go along with each one. In addition, being as much a rancher as a farmer, he greatly appreciated those early spring "weeds" that the livestock were finally able to graze on after a long winter penned up in a feeding lot eating nothing but dried remnants of last year's flora.

Henbit, chickweed, lambsquarter, dandelion and cheat grass. These were often the first ones to show up in the early spring and dad talked about when he was a child and those pioneers before him going out to harvest the tender lambsquarter and dandelion leaves to make the first green salad of the spring. He said after a long, cold winter, your body was craving the vitamins and minerals that could only be found in leafy, green vegetables, so they would eagerly eat these salads that would probably be considered rather bitter by today's more discerning palate. The henbit was always my childhood favorite, and it seems like it was of some of the other kids at St. Patrick's Elementary School as we would sit out on the hill during recess and suck the sweet nectar out of the purple flowers of this "weed." I taught my kids this "trick" at a very young age and they took to incorrectly calling henbit "honeysuckle" because they liked to "suckle" the "honey" out of the sweet flowers. If you have never tried this, next time you walk past a group of them in the early spring, throw away your adult inhibitions, lean down and pull just the purple flower part, which is hollow, then suck on the "bottom" end of the flower and you will get a little taste of the sweet nectar. No wonder bees love this stuff! I do it every spring and it always takes me back to my childhood.

For my five sisters, one of their favorites was the dandelion, and they were "master jewelers" at making dandelion chains that stretched long enough to make the prettiest necklaces. This process is one that many are familiar with. After the yellow bloom of the dandelion comes the tall stem with the "snowball" on the end--also a fun "playtoy" of the youngest of farm kids. After blowing the seeds and watching them drift away in the Kansas wind, the remaining head was simply pulled off leaving a long, hollow stem that was smaller on one end than the other. Making a circle, the small end is inserted in the large end and this method is repeated time and time again to make as long of a chain as is needed for jewelry--perhaps a pretty gift for mom in the house.

My mom was always appreciative of flowers, even if they were just Kansas weeds, and whenever we would bring her a handful of dandelion and henbit, she would stop whatever she was doing, get into the special cupboard over the sink where the breakable vases were kept, fill them up with water, put the flowers in them and set them proudly in the middle of the kitchen table to be the centerpiece at our next family dinner.

Perhaps kids today who grow up with perfectly manicured lawns of grass (that their fathers may not even allow them to *play* on) will grow up just fine sitting indoors on an early spring day playing video games while texting their friends, but you will have to forgive me if you drive by my house, with an early spring lawn that is a patchwork of henbit, chickweed, dandelions and cheat grass and you see my kids sitting out there picking flowers and making a chain for their mama while they drink their "honeysuckle" and soak up some vitamin D from the early March, Kansas sun.

April 26, 1991

I have been through bigger and deadlier tornado days than this one, but this one sticks out in my mind more than any other. Perhaps it is because I was a young, 23 year-old meteorologist working my first job and felt like I could save the world with my excellent forecasting or severe weather coverage. Or perhaps this one just hit too close to home.

I was working the overnight shift at WeatherData, Inc. at the time which was a solo shift. We had been looking at this day for about five days now, knowing it could be big. We were coming off of the unusually quiet decade of the 1980s that were shattered by the violence of the Hesston and Emporia tornadoes of 1990, and everyone's senses seemed more "tuned" to the severe weather potential, wondering what kind of a spring it would be. Late March answered that question dramatically when a monster tornado tore through the Willowbrook Division of Hutchinson and the video of that tornado, taken from on top of a building in downtown Hutchinson is still burned in my mind, even though it was virtually forgotten after April 26th.

My morning relief was Rick Dittman, who was doing the morning weather for KSN and as I waited for him to come in, I updated forecasts for railroads and utilities and spent more time than usual coloring maps with my pencils trying to figure out if today was indeed going to be "the big one." As I looked at more and more parameters, I concluded that this was indeed going to be the big one and I had a feeling in the pit of my stomach like when you eat a cheese sandwich too fast and have nothing to drink with it. When Rick came in, the first thing he said was, "How does it look?" Trying to lighten the feeling in the pit of my stomach I simply answered, "It's been nice knowing you!" Little did I know that later that day, Rick would be reporting live on KSN as he watched a tornado destroy an elementary school less than a half of a mile from his house. I wished later that I hadn't said it.

After I got off my shift, I actually had a couple of days off scheduled and planned on going out to visit the folks and get some farm work done. I was still in my storm chasing days, so decided it would be a good day to combine the two trips and would go out and sit on the dry line in southwestern Kingman County until the storms decided to fire. Turns out, I left town a little too late, and a severe thunderstorm developed in the warm, juicy air mass well out ahead of the dryline and I had to take 21st street and drive somewhat over the speed limit to avoid large hail on my new car that I had bought straight out of college. I got behind that first storm and headed for my original destination somewhere near Zenda. I made it to just west of Cleveland on the beautiful Cleveland Ridge blacktop when I felt the dry line and saw the towers going up rapidly overhead. I sat and waited. I checked a farmer's rain gauge to see how much he had gotten from the earlier storm. I smelled the growing wheat. And sat. The dry line refused to move. Turns out, it was waiting for the upper-level storm system to kick it out, and when it finally arrived, I underestimated how fast the storms would be moving and they quickly outran me. As I got a few miles south of Kingman, I saw baseball-sized hail covering the ground with a thick hail-fog. The first and only time I have seen these two things combined in my life. I stopped at a farm house and asked if I could use their phone to call in a storm report, called it in and by the time I got back out to my car, the first tornado was already approaching Goddard.

On my entire time out that day, I didn't see another single chaser. Chasing was different back then. No cell phones. No laptops. No data in the field except the NOAA weather and AM radio. I would plot the surface observations that were read on the weather radio hourly and make a crude and incomplete surface map, but mostly it was just watching the sky, reading the signs, smelling the dry line and trying to outguess Mother Nature. If that same storm happened today, there would be no fewer than 200 storm chasers lining the blacktops of Kingman County, all with computer data and cell phones waiting to move as one big mass of GPS-plotted information. Each has its advantages and disadvantages, but I sometimes miss just being out there by myself reading the signs instead of a computer screen.

As the first tornado was hitting Goddard, I decided to head back to the safety of the farm and watch and listen to the coverage of the storms as I watched the impressive, massive cumulonimbus clouds billow up to my east.

An interesting story about the tornado that hit Goddard, then lifted near Saint Mark's: TV crews from Wichita heard that Goddard had been hit and they sent an awful lot of their resources in that direction to cover it before checking with their meteorologists. When the "Andover Tornado" began forming near Clearwater, most of the crews were out of place to capture what would be the really big tornado story of the day.

The tornado is well-documented in many books and was one of the first to be widely videotaped as hand-held VHS cameras were becoming commonplace among more than just the wealthy. I won't go into extensive analysis of the tornado here, but a few things about it that many may not know.

The national lightning detection system was in place by that time and we had a tap of it in WeatherData. As we went back over the data of the day, something startling was discovered: in the lifetime of the supercell that produced the Andover tornado, there was almost no cloud to ground (or ground to cloud) lightning. It is a mystery to this day as to why.

The famous video of the tornado as it rapidly intensified going across McConnel Air Force Base has become something of legend. The videotape was supposedly thrown over a fence on the edge of the base to a waiting reporter who made a copy of it then gave the original back. The man who shot the video was supposedly threatened with a court martial for shooting video on a classified Air Force Base and releasing it to the media. He and the original videotape have essentially disappeared, and the best copy of that tape known today is still only second generation.

The third tornado in what would become a family of tornadoes moved up the Kansas Turnpike on the east side of El Dorado

Lake and a KSN reporter (Greg Jarrett) and photographer (Ted Lewis) were on their way back from Topeka where they were doing a special story on an Indian Reservation near there. They were unaware that they were driving into a tornado-producing supercell, but were tipped off to the fact when pink, fiberglass insulation from Haysville, south Wichita and Andover began raining from the sky. As they saw the tornado off to the east side of the turnpike, they decided to try and get ahead of it so they could shoot good pictures of it. Had they kept driving, they would have been in no danger at all. The tornado quickly overtook them, and by the time they reached the first overpass northeast of El Dorado Lake, they had to make a split-second decision that remains controversial to this day. They decided the safest place for them to be would be up under the steel girders of the bridge instead of lying flat in the broad, rain-filled ditches of the Kansas Turnpike. The video they shot from under that bridge was some of the closest to the inside of a tornado that has ever been shot and the section that shows the minivan being bounced down the ditch like a basketball still haunts me to this day. After that video, unfortunately, many mistakenly thought that under an overpass was the safest place to be in a tornado, and there are documented cases of where people left the safety of their homes to drive to an overpass to take cover from an approaching tornado, and some have died doing so. Even today, the myth persists that under an overpass is a safe place to be in a tornado, which, of course, it is not.

After the tornado was over and lives were lost and shattered forever, it was a tough thing for a young meteorologist to see, and I worked in an office full of young meteorologists. We had tears and anger and meetings to talk about our feelings; "Could we have done anything more to save the lives of those lost?" "Why didn't the man out walking his dog take cover?" "What can we do better in the future?" As the funerals started taking place and the debris started to be cleaned up, we had to live with and drive through the area that had been so heavily damaged, and it was the first time in my career that I remember learning that it is all about the people and not the tornadoes. Unfortunately, it was a lesson that I would be taught again and again through the rest of my career. I see some of the "young kids" that love to chase for fun and they are so excited to see a tornado that they are literally rooting for one, but after April 26, 1991, I never rooted for another tornado; I always rooted for the people.

When Weather Kills the Ones You Love

Growing up in Kansas in a big, Catholic, farm family means you are related to a lot of people in the state. I was one of seven children, my father was one of seven and my grandfather was one of seven. You can do the math, but that means that I have dozens of first cousins and probably somewhere in the neighborhood of one hundred second cousins, many of which live in Kansas. Most of my cousins on my dad's side grew up in the Dodge City area and those on my mom's side mostly grew up around Kingman. I was fortunate as a child to get to know most of my first cousins pretty well and got to play with them and make friends with them when I was young.

This can leave me especially anxious in times of tracking storms, though, especially when towns where my cousins live pop up on a storm track from a particularly nasty-looking storm: Spearville, Meade, Garden City, Hutchinson, St. John. I worry about my relatives in the path and say a quick prayer that they might be kept safe from the approaching storm.

On May 25, 2011, the unthinkable happened. I was backing Dave Freeman up on severe weather coverage and there were a lot of severe storms tracking across the state, some which had "embedded" tornadoes in them that were completely wrapped in rain, hard to see and tough to pick out on radar. One such storm moved between St. John and Great Bend and I remember thinking at the time how lucky we were that it was missing the town of St. John where my cousins were. What I didn't know at the time was that they were in a vehicle traveling south from a shopping trip in Great Bend to pick out furniture for their newly-graduated daughter. As they drove, the storm got worse and they decided to pull off of the highway and take refuge in a private driveway until visibility improved enough for them to proceed on to their home in St. John. In a case of "doing the right thing but simply being in the wrong place at the wrong time," a rain-wrapped tornado moved right across the farm that they had pulled into to take refuge. It uprooted a large Cottonwood tree which became airborne and was dropped right on their vehicle killing my cousin's wife and son and seriously injuring their daughter.

The report of the fatality came into the WeatherCenter and our hearts sank but I still had no idea who it was that had lost their lives. Soon, the cell phone in my pocket started ringing. I was too busy to answer (we were in continuous coverage of the storms on KSN), but I looked and it was my aunt calling me, which I thought was unusual. Then it rang again. My sister. And again. My wife. Then a text from my sister. Something to the effect of "you need to call me immediately. family emergency." I don't remember much of what happened next and I don't remember even how I got home that night, but after calling her back and hearing the news, I was no longer able to function at work and was dismissed.

The next days were a strange mixture of celebration and sorrow. We had been working for six months on our farm's 100th anniversary since the Sheahan family had come from Iowa and moved to the farm in the spring of 1911, a party we called "Sheahan-tennial" so it was decided that the party should go on. My deceased cousins were planning on attending and the party just wasn't the same without any of the cousins from that family there. The rosary was the day after the party and the funeral the day after that. Funerals are never enjoyable, but when the deaths are of young, healthy people that die in a sudden, tragic disaster in the prime of their lives, it is especially hard.

I suppose it is a hazard of the profession I have chosen and the fact that I have stayed in my home state my entire career, but after the loss of my dear, loved ones, that sick feeling in the pit of my stomach when one of their towns shows up in a storm track is even harder to push through. Also a sobering reminder that in the track of *every* storm is someone's loved one that would be terribly missed if they were taken away by a Kansas storm and I pray that if they have chosen to watch and listen to me for storm coverage, I can say the right things at the right time to protect them from harm.

The "Magic" of Late May

Something almost magical happens to weather in Kansas once mid-May rolls around. It is like the storms are supercharged and it is almost impossible to get just a plain, old "generic" thunderstorm. Every little cell seems to become severe and produce large hail or tornadoes. Even on days where the setup in the atmosphere looks marginal by all measures, a storm can fire out of nowhere and produce impressively significant severe weather. A day that looks like it will just produce a severe squall line will suddenly morph into storms that create a dozen tornadoes. A day where you think there is no way that the moisture is deep enough or the front is strong enough or the winds are too light that you have completely written off can suddenly turn into a busy evening of tracking a stubborn cell that refuses to die.

Of course this late-May "magic" can become magic of the "dark" variety and the consequences can be deadly. If you look at the history of Kansas tornadoes, some of the biggest and worst have been in late May, including the deadliest tornado in state history, the tornado that struck Udall on May 25, 1955 killing 80.

Of course, Memorial Day weekend always falls in late May as well and being the "unofficial start of summer" brings many people to the area lakes. It is said that on Memorial Day weekend, El Dorado State Park becomes the 3rd largest city in Kansas and I haven't seen a tent or camper yet that came with a basement, so it leaves a lot of people vulnerable.

Late May is also when the wheat fields of Kansas are turning their amber color to make the grain that will be baked into bread for the entire world for the next year and the brittle straw and heads of wheat become especially susceptible to even the smallest hail, especially if it is being driven by the wind from a Kansas thunderstorm.

It is also the time that kids are getting out of school, graduations are being celebrated and it seems that every single day has something on the calendar.

All of this comes together to give this time of the year an almost electric feeling of the busiest time of the year, the stormiest time of the year, the most fun time of the year and the most anxious time of the year all rolled into one. It almost makes you look forward to the long, boring, dry, dog-days of summer that are only a month and a half off. Almost.

Working with Legends: Mike Smith

I have had the fortunate opportunity in my career to work with several people who have made such an impression on the weather community that their names and contributions will be remembered and felt long after they are gone. One is Mike Smith. Mike gave me my first job straight out of college when he was the Chief Meteorologist of KSNW-TV and the owner and president of WeatherData, Inc. (later bought out by AccuWeather).

Mike developed several new methods of weather analysis and severe weather warnings and was often thinking up ideas that were ahead of the technology that was even possible at the time. The one that made the biggest impression on me was the digital, colorized radar. KSN and WeatherData had purchased an old WSR-74C radar. The WSR stands for Weather Service Radar, the 74 stands for the year it was designed (1974) and the C stands for the radio band that the radar uses. This is the radar that replaced/supplemented the WSR-57 radars that the National Weather Service used up until the WSR-88D (the D stands for Doppler), commonly known as NEXRAD, which stood for NEXt generation RADar. The problem with the 74 and 57 radars was that they weren't very easy to view and really didn't make for very good TV. The display on them was a round, phosphor screen like something you would see in a submarine, and you needed it to be fairly dark to see the echoes very well as they were "painted on" by the sweep of the radar, only to fade to black before being "painted on" by the next sweep. The only way to look at a "loop" of the radar was to mount a camera over this display and view it in a time-lapse recording. The other way was to use grease pencils and draw an outline of an echo, wait 10 minutes or so, then draw an outline of the latest echo and use extrapolation to determine how fast and in what direction the storm was moving. Mike contacted someone he knew at a company in California to design a better way to display this and one that would be much more pleasant to view on a color TV. The man needed to know what colors to make the echoes for the display, and Mike told them he thought that light rain "felt like" it needed to be a green color, moderate rain a yellow color and heavy rain and hail a red color. This basic color table is still in use by almost every radar display in the world. The man came up with an ingenious box that took the information out of the radar and did just that and plotted it over 3 custom maps at 3 different radar ranges. This new, revolutionary, colorized radar display was used on the air at KSN up until the early 90s by which time, others had jumped on the bandwagon, digitizing the radar information from every National Weather Service office in the country. I had the opportunity to look inside the "magic box" that had been created for this purpose several times, and it was a work of electronic art. The entire circuitry of it was hand-soldered, wire-by-wire, neatly laid out and was really something to see. Unfortunately, when the WSR-74C was sold for junk/parts in the early 90s to a company in California, that box was sold with it and most likely ended up in a scrap pile somewhere. It should have ended up in a broadcast or electronics museum or at the very least, a museum somewhere in Wichita. I was still working at WeatherData at that time, subcontracting out to KSN on the weekends and before I realized what was happening, it was gone and I missed my opportunity to save that piece of weather history.

Mike was also personal friends with some of the other pioneers of meteorology and brought them in to talk to our staff about weather. The most memorable of these was when he brought Ted Fujita in after the Hesston tornado. If that name sounds familiar to you, it is because he is the man that the tornado intensity scale is named after. It was called the "F" scale for many years, an open-ended scale in which tornadoes fell into the F0 to F5 categories. Later on, the "Enhanced Fujita" or EF scale was devised and it capped things off at the EF-5 level. Dr. Fujita was getting very late in life when I heard him speak and still had his thick, Japanese accent. In addition, he spoke softly, so you really had to listen and pay attention to catch what he was saying. I will never forget that he said he believed the Hesston tornado was the most powerful tornado to strike the face of the earth in recorded history and he believed there might have been some F6 winds that occurred, but he could not say for certain without a Doppler measurement, since there is no "damage measurement" for winds that strong. In other words, once the damage reaches the F5 category, it is so horrific and complete, that there are no "markers" of what F6 winds could do above and beyond the total destruction of the F5 winds. He challenged those of us meteorologists in attendance that day (in what is now the KSN newsroom) to be the first to find that legendary F6 tornado because he believed it existed and believed that it would probably occur somewhere in Kansas or Oklahoma. His prophecy was almost fulfilled on May 3, 1999, the day of the F4 Haysville tornado when the Moore (Oklahoma) tornado had winds measured by Doppler radar just under the 318 mph required to put it into that "legendary" category. The Moore Tornado would be the last F5 to hit the U.S. before the new "EF" scale would be put into official use, so no F6 was ever found. The Greensburg tornado was the first EF-5.

Mike Smith continues to be on the leading edge of meteorological breakthroughs to this day and we remain friends and have lunch together on occasion. I recently had the opportunity to interview Mike about his second book that he has written. His first book is an historical look at his career and "How Science Tamed the Weather" and his second is a controversial and raw look at what went wrong in the deadly EF-5 Joplin (Missouri) tornado. After the interview, in which the student was interviewing the former teacher, in the station and the very room and in the exact spot where the old WSR-74C stood, I told him what a great job he did and that he "just might have a future in this business!" We shared a good laugh.

Tornado Detection: 1970s Kansas Farmer Method

I grew up on a farm in western Kingman County, about 60 miles from the nearest TV or radio tower. A tornado siren was something I heard my "town friends" talk about, and being on rural electricity, about any time the wind blew or lightning flashed, the lights would flicker or fail. FM radio was pretty much non-existent before 1980 in this part of the country and the four TV stations that came in by antenna would fade in and out and were especially hard to get after sundown. The only AM radio we got after sundown was from Chicago or Des Moines or Dallas or Denver. This meant that you were pretty much on your own when it came to detecting and taking cover from tornadoes.

My parents' approach to this with us seven kids was to pretty much go to the basement or cellar any time they "had the feeling" we needed to. Fortunately, they erred on the side of caution, but unfortunately, that meant spending an awful lot of time under ground as a child. My dad had heard about different ways to "detect" a tornado, one of which was to tune your TV to the lowest channel, channel 2, and the lightning within a tornado was supposed to make a distinct pattern on the screen. I can remember getting up at night during a storm and seeing dad sitting in front of a TV full of static, trying to look for whatever that pattern was. Of course, if a storm was especially electrical, the static on the TV would jump with every stroke for who-knows-how-many miles, and if it reached past my dad's comfort level, we would head for the basement one more time.

Going to the basement that often meant being prepared at all times. We had bags packed at the foot of our beds with some extra clothes and perhaps some of our most-loved toys. I remember my sisters having a separate suitcase just for their Barbies. We would leave our shoes at the sides of our beds when we went to sleep every night so that we wouldn't have to hunt for them should a "midnight tornado" come up. Looking back, my parents were way ahead of their time when it came to disaster preparedness. This is some of the same advice that we still give today.

Being a Catholic family, going to the basement also meant a lot of prayers that the storms would pass us over. We would usually stay down in shelter long enough to pray a rosary, light a holy candle, perhaps carry some blessed palm branches from the recent Palm Sunday and always the Holy Bible. A lot of those nights are blurs to me as I am sure most of the seven of us kids drifted in and out of sleep while mom led the rosary and dad snuck up the stairs to take a look out the porch window to see if the "all clear" could be sounded for us to go back up and go to bed. One time, dad went up only to see a tornado in the field less than a quarter of a mile from our house in a flash of lightning. He came running back down and luckily, the tornado lifted at the fence line before touching down northeast of our house again.

In this day and age of blame; "Why didn't the sirens go off?" "Why didn't my weather radio sound?" "Why didn't the warning show up on my smart phone?" perhaps a lesson could be learned about personal responsibility from the way Kansas farmers have been doing it for over 100 years. Perhaps, when the Kansas sky grows dark and stormy, the calendar says "May" and you just get that feeling that something isn't right, we shouldn't wait for someone to "tell" us to go to shelter. Perhaps we should just grab the ones we love and take them to the basement. A few prayers probably wouldn't hurt anything either.

May in Kansas

Growing up, May was my favorite month. Now that I have become an adult, I rank it as my second favorite (after October).

As a kid, May meant many things, not the least of which was the end of school. I wasn't a bad student and didn't necessarily hate school, but I sure preferred being at home and on the farm and May is when things really started to get exciting.

Sometime in early May, my dad would decide that it was time to start working on the combines to get ready for wheat harvest. We were what one would call "junk farmers," and drove very old machines that sat outdoors all winter without a machine shed over them, so they would get quite rusty, and just the task of starting them up, airing up the flat tires and moving them to the shop to begin the work to get them ready took the better part of a day some years. My heart thrilled at the sound of the engine when it finally did fire after many attempts with the jumper cables, perhaps replacing the fuel pump, spraying ether into the carburetor and cranking and cranking and cranking. There is no other sound in the world like the sound of an older-model, gasoline-powered combine due to the fact that the motor sits out in the open and high up on the back, and whenever any of the neighbors within a couple of miles fired theirs up, you could tell it wasn't a truck or a tractor or anything else. It was unmistakably a combine.

Driving a combine is perhaps my favorite thing in the entire world. I dream about it year-round even though it has been years since I have actually driven one. Perhaps my second favorite thing is working on them. After we would finally get the combines running and moved to the shop, it was time to oil all of the chains and the sickle, grease all of the bearings and zirks (the old combines we used had close to 50), tighten all of the belts, fix all of the rust holes, clean out all of the old grain from last year that had rotted in the machine, change the oil and get everything working the way it should.

We drove old John Deere combines made primarily in the 60s and my dad would always pick up a couple of cans of John Deere Green paint when he bought parts for the first time in May. It was my job to make sure that the new parts we put on were shiny and freshly painted. It sounds silly doing this to such old machines that just sat out in the weather all year, but it was a sign of optimism and pride. A sign that the crop would be good. Good enough to pay for frivolous things like cans of paint.

I spent most of my days in school in May daydreaming about coming home to the farm to rush across the road to the shop to start working on the combines. We would work until mom called us for a late supper or until a powerful, May thunderstorm came up and drove us inside.

May has the most violent weather of any month in Kansas and the combination of that, the end of school, the anticipation of harvest and the warmer weather was about as close to heaven as a young meteorologist, farm-kid could find.

I remember the sickening feeling of especially violent thunderstorms that would bring a green tint to the sky and rain hail down on the wheat, wiping out in minutes what it had taken nine months to grow. One year, on the last day of school, it was my oldest sister's birthday and we came home from our "last day of school picnics" to have a birthday dinner. The sky grew dark and we started hearing banging noises from outside. The entire family rushed to the front porch to watch helplessly as the hail got larger and larger until it began shattering the windows on our vehicles. It had rained the day before and there were large puddles sitting around and I remember thinking it looked like a war movie when the baseball-sized chunks of ice would hit the puddles sending water flying into the air as if though a grenade had gone off. We went ahead and cut the wheat that year, as we did every year, but there wasn't as much grain to haul to the elevator thanks to those nightmarish chunks of ice from the sky.

As an adult and a professional meteorologist, May now means long days without dinner breaks followed by short nights and more long days covering storms. While I still thrill at this to a certain extent, I often find myself longing for that optimistic and proud smell of oil, grease, wet straw and John Deere Green paint fresh out of the can.

SUMMER

June 19th...Violent (and Expensive) Weather Days in Kansas

In my first three years as a meteorologist I had the chance to witness and work two historic days of severe weather that just happened to both fall on June 19th.

The first one occurred in 1990 in a storm that came to be known as the "Inland Hurricane." It was a hot June day at the height of wheat harvest and Mike Smith was off on a business trip, so a young, inexperienced Ken Smith was assigned the primary broadcasting duties on KSN that day. I remember him saying at the beginning of his shift that he hoped it stayed quiet as he had never done a severe weather interrupt before. That would soon change. I was working the "Plains Shift" as it was called at WeatherData, Inc. that day so I would be forecasting and supporting Ken on his duties. All signs were pointing to it being an "all or nothing" kind of a day where if the cap held, there wouldn't be a single storm, but if it broke, the storms would likely become severe very rapidly and with surface temperatures approaching 100 degrees, they would likely have a lot of wind with them. I worked ahead on my shift that day in case anything happened and was done ahead of time. Ken and I were remarking that maybe nothing would happen after all, but it was only approaching 6PM at that time and we knew there was plenty of daylight left for a storm. Sure enough, a small "blip" showed up on the radar in Pratt county that grew rapidly with every scan of the 74C radar that we had on-site and a warning was soon issued for this lone severe thunderstorm. Watching it on radar was almost like watching something organic. Every sweep it grew and grew and started to take on the signature "bow" shape on the front with a tight leading-edge gradient from no rain at all to the bright reds of the radar over a matter of less than a mile--both classic signs of strong winds.

My family was harvesting wheat near the town of Calista when it hit there and they were trying to rush the truckload to the elevator before the storm hit. They saw it go up and claimed that shortly after it went up, it moved across a field of burning wheat stubble and they could see the flames and smoke first getting sucked up into the storm, then blown out violently ahead of it, spreading the fire to neighboring fields that hadn't been cut yet. They barely got the truck of wheat unloaded at the elevator and were on their way back to the farm when the storm hit. Gustnadoes danced in the fields around them and one came very close to hitting the truck as they raced down the road.

This was in the days before cell phones, so the first word we got of major damage at KSN was that large, old trees were being uprooted in Kingman with winds estimated at 100 m.p.h. It was clear then, that we had a "Derecho" on our hands. A Derecho is a long-lived wind storm that can travel for many miles producing widespread wind damage. The line of storms continued to lengthen and strengthen and by the time it got to Sedgwick County, it was doing major damage from Colwich and Valley Center all across Wichita. Every TV and radio station was blown off the air as towers collapsed in the up to 120 mph winds and trees were shredded. Electricity was out for most of the cities in the storm's path and the storm just kept going and going. It produced widespread wind damage across an ever-widening path all of the way to Emporia and into western Missouri and became the costliest storm in Kansas Gas & Electric's history blowing down miles and miles of high-tension power lines through the Flint Hills and beyond. When it was over, $80,000,000 worth of damage had been done and 33 people had been injured.

After getting done helping with the severe weather coverage far beyond my normal shift time, my brother and best friend from Kingman came into town to view the damage. They were the ones who had seen the gustnadoes near Calista early on in the storm's evolution. It was strange as we drove around town because there were no lights anywhere, trees down everywhere, and the "scan" button on the radio would just go round and round the dial since there were no stations on the air to stop at. Finally, a new station at the time, B-98 who had studios and a tower downtown were able to get back on the air (I assume at low power) and they were the first source of information anyone had on how bad the storm had been. Most TV and radio stations were off the air until the next day or later.

We had heard that the high-tension power lines near the Colwich KG&E plant had been blown down, and since my brother had worked there in college, we wanted to drive up and see. As we drove up K-96 toward Colwich, the road was closed due to the fact that there was a barn sitting on its roof in the middle of the highway...a sight I will never forget!

Ironically, exactly 2 years later a different kind of violent Kansas thunderstorm would strike but this time it would be a young, inexperienced Mark Bogner at the helm who was hoping as his early-morning/noon shift began that he wouldn't have to do an interrupt because he had never done one before either.

I had a feeling it was going to be a bad morning when I drove down Kellogg on my way to work and saw something I had never seen before and have never since. The sky was dotted with a certain kind of cloud called "Accas," short for altocumulus castellanus, a sign of strong warm-air advection about 10,000 feet off the ground, and the Accas actually had lightning flashing in it! To me, that was a pretty good sign that storms were going to form very soon and very close to the Wichita area, and sure enough, severe storms developed and even produced a tornado near Andale. The cut-in for the Andale tornado *would* have been my first severe weather interrupt, but Mike Smith had seen what was happening and rushed into work still in his pajamas, straight out of bed. This is the famous "Mickey Mouse T-Shirt" incident where Mike did tornado interrupts in bed-head hair wearing an old t-shirt. For the record, it was actually a "Bowling for Big Brothers/Big Sisters" t-shirt, but the "Mickey Mouse" thing sounded funnier and stuck. The reason that Mike did the interrupt instead of me was that one of our clients at the time was the Texas Ag Network. Mike had never done the TAN before and I had never done a TV interrupt before, so it was decided that we would do the roles we were more familiar with. The TAN contract had us doing agriculture-specific forecasts for every county and zone in the state of Texas and took a good 45 minutes to record some days because of all of the micro-climates that Texas has. While I was recording, the storm that produced a tornado near Andale moved southeast right over downtown Wichita and the KSN studios. I was in a sound-proof radio booth, but golfball hail began falling so heavily and for such a long period of time that I had to shout to be heard over it to do the recordings for TAN. In some parts of the city, the hail got as large as softballs. The client called to complain about the quality of recordings that day, but were more forgiving after they heard what we were going through.

Unfortunately, that early-morning wake-up call was just the first shot across the bow that mother nature would fire and as mid-morning rolled around, a particularly strong thunderstorm developed over the city of Russell and I finally *did* get to do my first interrupt of programming for baseball-sized hail. The storm continued to roll south-southeast and as I was getting ready to go on for the noon show, the storm slammed into Hutchinson with softball-sized hail and 70 mph winds. I had been doing frequent interruptions for this storm as it continued to progress and as it moved south, it was accelerating and not losing any strength. It was what my dad would have called a "northerner" much like the storm that struck when I was a child that made me want to become a meteorologist in the first place. It was also becoming frighteningly apparent that the west side of Wichita was going to take a direct hit right about the time the noon show was getting over. Sure enough, the storm hit northwest Wichita with the same softball-sized hail accompanied and driven by 70 mph winds. The damage along West Street in Wichita was catastrophic as nearly every window and sign was blown out and shattered. Also hard-hit was the Twin Lakes area at 21st and Amidon where hail drifts were still found 12 hours later, Riverside Hospital where people were injured by hail trying to get from their cars into the hospital, Kansas Newman and Friends University, where every window on the north side of the buildings was shattered. People out driving in parts of west Wichita had every single window blown out of their vehicles along with body panels actually being punctured by the hail. There were several stories of people who had had their windows replaced from the early morning storm only to have them blown out *again* by the second storm. The damage to homes in west Wichita was way more than to roofs and windows. Some houses had all of the siding stripped off the north sides! When the storm finally ended, 59 people had been injured and $650,000,000 worth of damage had been done making it the costliest natural disaster in state history up to that time.

For some reason, probably because I was too busy, I didn't have a chance to go out and look at my car parked at KSN until my shift ended that afternoon and sure enough, it was pretty much totaled by the storm. I never did know if I got more damage from the early-morning storm or the mid-day one.

One thing I do know, whenever the calendar rolls around to June 19th every year, I get a funny feeling in my stomach and watch the sky just a little closer than usual.

Just What the Heck is the Dew Point Anyway and Why Should I Care?

Believe it or not, there is a much more accurate and descriptive way to talk about the moisture in the air and this is the dew point. Often, relative humidity is what is used, but there are many problems with relative humidity, most important that it is *relative* to the temperature at the time. As a result, the number in and of itself means nothing unless you know the temperature. For example, if I told you that the relative humidity today was 35% would you think that was a high humidity or a low humidity? Most would off-the-cuff answer "low" because it is somewhere below 50%, but remember that it all depends on the *temperature* at the time. If it is 105 outside, a 35% relative humidity is not only very high, but uncomfortably so. On the other hand, if it is 15 degrees outside, a 35% relative humidity is so dry that your lips and skin are likely to crack and chap due to the extreme dryness.

The other problem with relative humidity is that it is "diurnal." That is, it changes from hour to hour and is usually the highest in the morning and lowest in the afternoon, so the number could theoretically vary from 100% to 1% within the same day!

The dew point temperature is superior in several ways. It does not have that "diurnal" factor nearly as much as the relative humidity does and is much more able to be correlated to a comfort factor since a 45 degree dew point temperature will bring about a similar reaction from your body no matter what the temperature is outside.

First of all, what exactly *is* the dew point temperature? It is simply the temperature that the air would have to be cooled to in order for dew (or frost if it is below 32) to form. For example, if it is 75 degrees outside with a dew point temperature of 65, you would have to cool the air to 65 degrees in order for dew to form. Pretty simple.

In addition, there are some warm-season "rules of thumb" that can apply to the dew point temperature that can help you decide what kinds of weather might be in store.

Dew Point	Weather	Comfort Level
30s or lower	Hard to even get a thunderstorm	Dry with lots of static electricity
40s	Severe weather unlikely	Quite comfortable
50s	Thunderstorms possible but not likely severe	A little humid-feeling
60s	Any thunderstorms that develop will have a fairly high chance of being severe	Uncomfortably humid
70s	Violent storms are possible	Extremely uncomfortable, "hard to breathe" air
80s	Deeply Tropical	Danger! Very unusual in Kansas

As with any rule of thumb, there will always be notable exceptions, especially in the high plains near the Colorado line. Here, severe storms can get going with dew point temperatures in the 30s!

So, now that you understand the dew point temperature a little more, consider how it is possible that a 35% relative humidity can be either extremely uncomfortable, comfortable or extremely dry.

Relative Humidity	Temperature	Dew Point	Feels
35%	75	45	Comfortable
35%	105	72	Extremely uncomfortable, "hard to breathe"
35%	15	-8	So dry, things crackle

If it were up to me, we would stop reporting the relative humidity and just do a much better job of educating the weather audience on what the dew point meant and use that, but even though it is quite misunderstood, people still love the relative humidity and still want it included in their regular weathercast.

On an interesting aside, the highest dew point ever recorded in the United States is *not* along the Gulf Coast, but in the middle of corn fields. When corn is growing in July, it emits or "evapotranspirates" an incredible amount of water--an acre of corn gives off as much as 4,000 gallons per day! On July 13, 1995, Appleton, WI measured a dew point temperature of 90 degrees! The world record, however *is* near water, where water temperatures as high as 98 have been recorded (also the hottest in the world). Dhahran, Saudi Arabia recorded a 95 degree dew point temperature at 3 p.m. on July 8, 2003. I can't even imagine how uncomfortable and deadly that would be. Thank goodness we will never have to find out in Kansas!

When a Season Should be Over Already

It seems like it happens every season, but especially Summer and Winter, the two longest seasons in Kansas. Just about the time you think it is over, the season, like an army that is just about defeated, puts together one last "all or nothing" push trying to spend its last energy on one final assault on the tired senses of the people defending themselves from the heat or cold. The winter push usually comes in March after a couple of beautiful 70 degree days that have fooled everyone into thinking winter is over. In summer, this usually hits in September after several days with highs in the 70s and some crisp mornings. Your senses are just convinced that the back of summer has been broken and the heat is gone for good. Then it hits. Windows get closed up again and air conditioners that have had a chance to breathe are called back into service. Pools have been long-closed and lake water is getting too cold to spend much time in, so you just have to ride out that hot week. People are in more of a bad mood than usual, as people who have been wronged in some way; promised one thing but given another; a snake for an egg.

If you can get past the indignation, there are many signs, of course, that this last feeble push will ultimately fail. I am blessed to live in a house in the city that is covered by a giant Oak tree of beautiful stature. It is wise enough to have seen many false seasons come and go and knows that time is constantly marching forward, so long before the days are cool to stay, when temperatures surge up to near 100 again, it gets on with the business of ensuring propagation by dropping enough seeds to cover my back yard and fill my gutters. The little caps of under-developed acorns come first but as the days get shorter, the acorns grow in size. First, big, green acorns fall, then come the fully-developed, fat, capped brown ones that you see in children's alphabet books under the letter "A."

As I sit here even now in the closed-up, air conditioned house with the sun beating down and the thermometer looking more like July than September, I can hear that fall's cooler weather is indeed right around the corner as the acorns fall like hail on the roof making a distinct tap-tapping sound over my head. My favorite season is within reach and I close my eyes and smile.

A Boredom for Every Season

It may sound funny, but EVERY season in Kansas reaches that time where you run out of ways to say it. In the spring, you get tired of having a chance of thunderstorms in every day. In fall (believe it or not) it gets boring to talk about highs in the 70s and lows in the 50s until further notice. Winter blahs come from those days where you don't have any winter storms in the forecast, but you don't have any days that are either warm or remarkably cold.

The last two summers, however, have been the most challenging of all. Not only is there nothing new to report in the weather for weeks on end, but it is always news that nobody wants to hear, including the weatherman himself. People often forget that we, too, have lawns, gardens, farming interests, etc. Since forecasting any major change beyond about 5-7 days is beyond our current level of forecasting skill, questions like, "when is this heat ever going to break?" are often met with a sarcastic answer like, "September."

The thing that keeps it interesting to me is the statistical part of it. Being a native Kansan and loving this state so much, I love it when we get to challenge any kind of long-standing record. I honestly didn't think I would ever see some of the records that were set in 1936, but have gotten to see several of them fall just in the last 2 years, including the incredible numbers of 100+ degree days (53 in Wichita in the summer of 2011, breaking the long-standing record of 50 from 1936, for example).

Fortunately, just about the time that you can't think you can take another day of the monotony, Kansas throws you a curve (like a cool, rainy 70 degree day in the middle of August) and another entirely different season is suddenly just around the corner with all of its excitement...at least until the next "boring" streak kicks in.

A Kansas Summer Danger: Heat Stroke

One of the many dangers to working on a farm is injury or illness due to the weather. From frostbite in the winter to heat stroke in the summer, there is plenty a person needs to protect themselves from while laboring outdoors. One of the biggest dangers leading to heat stroke, in my experience, is putting up hay in July and August. Not only are you out in the middle of a field where there is no shade, under a sun that is beating down, but the temperatures often exceed 100 degrees, and that is the temperature in the *shade*. Step out into the sun to "buck" 70 pound bales and you can easily add 30-40 degrees to the *shade* temperature. It is a bad combination: extreme physical labor and extreme heat. A lot of farmers and ranchers choose to do this kind of work either very early or very late in the day, trying to take advantage of the cooler temperatures and less-direct sun angle, and that helps some, but the humidity is often higher during those times of day causing more stress on the body.

My heat stroke experience came at just such a time when I was a teenager in the early 80s. We had loaded a trailer with hay as the sun was setting one summer evening and pulled it to the barn to be unloaded. Since it was dark and we were tired, it was decided that I would arrive before sunrise the next morning and unload the trailer myself. This was at my uncle's farm, where I worked throughout high school helping check and raise registered Hereford cattle and was about 10 miles from our home, so I set out early that morning so I could "beat the heat." Unfortunately, it was one of those summer mornings where the low temperature didn't even drop below 80, there was no wind, and dew point temperatures were in the 70s, bringing extremely dangerous conditions in which to do manual labor.

The barn where I was unloading the hay was actually an old, stone, converted milkhouse from the days where every farm had a milkhouse. A garage door had been added to the southeast corner and there was only one window in the entire building which also faced south, so it was impossible to get any kind of a cross-breeze.

When you put up hay, there are several requirements in clothing that don't help the heat situation either. You must wear jeans and it helps if you have leather chaps on over them to protect your thighs from cuts and scratches. You must also wear a long sleeve shirt (usually over a t-shirt) to protect your forearms from cuts and scratches. "Bucking" a bale is made up of a continual motion where you grab the bale with both hands by its two strings, stand erect to get the bale off of the ground, "kick" the bale up to your chest level with your right thigh, bounce it against both of your forearms and then use the momentum you have gained to throw the bale as close to where you want to stack it as possible. It really builds muscle, but is very hard on the body.

This particular morning, I was dressed for the part and my teen-age logic was that the later I waited to do this, the worse the heat was going to get, so I went at it like I was, "killing snakes." I would throw about 10 bales down off of the trailer, then would have to go inside the stiflingly hot barn and "buck" those bales one more time to get them nicely stacked in the barn so that the maximum amount would fit in and the stack wouldn't fall down as you removed bales in the wintertime.

Needless to say, I was way overdoing myself, not taking nearly enough water breaks and just trying to get it done as quickly as possible and by the time I got the last bale unloaded, I was physically in trouble. Before I even started stacking the last batch, I started getting chills and stopped sweating, but pushed through it to get it done. I remember coming out the door of the barn, going to my knees and throwing up. I didn't have the strength to get back on my feet, so I crawled over to the car that I had driven in, pulled myself up into the driver's seat and drove home via dirt roads. I may have even blanked out once or twice because I don't remember anything about the drive home except that now the heat actually felt *good* as my fever continued to rise and the chills became worse.

By the time I stumbled into the house, I had a pounding headache, a raging fever and was throwing up again. My mom recognized that something was wrong and she wisely poured me a cool bath, and I sat in there shivering and moaning, but it probably saved my life by lowering my out-of-control core body temperature. In hindsight, I should have gone to the hospital, but instead just stayed at home and in bed for several days with headaches and fever coming and going. There is little doubt that I had either extreme heat exhaustion or heat stroke (also sometimes called sunstroke).

My body temperature regulator has never been the same since that hot summer morning and to this day, if I start to overdo it even a little in the heat, I get the same chills and feeling of exhaustion, but now I am smart enough to know that it is my body trying to tell me to shut it down, go inside and sit under a fan in the AC and drink a tall, cool beverage.

The Long, Hot, Boring, Dog-days of Summer

There were few things more exciting as a kid growing up in Kansas than the first day of summer. The mind was filled with lists of things that would get accomplished over the next 90 days without having to worry about school and the duties and time-robbing that went with it.

Growing up on a farm, the first month was the most fun. It was filled with lots of duties to get through wheat harvest and putting up hay and getting cattle worked, but you always knew that even when harvest was done, there were *still* 60 days left.

Then reality hit. My dad always said, "Rain follows the harvest and drought follows the plow." His theory was that in a matter of a couple of weeks, the landscape went from green, ripening wheat to golden waves that seemed to bring with them waves of severe thunderstorms, to turning the soil over with a plow and disk, making it a dark, hard, dry brown that just seemed to rob the moisture from the air and make the wind blowing across it feel like it was coming from a desert.

I later learned in microclimatology classes at KU that there was a lot to my dad's theory. Green wheat undergoes what is called "evapotranspiration" whereby it pumps millions of gallons of water into the lowest, "boundary layer" of the atmosphere. As the wheat ripens, the landscape's "Albedo" changes drastically and like a cover of fresh, white snow, a larger portion of the sun's energy is "bounced back" into the atmosphere, feeding storms. Then, very quickly and very dramatically the "Albedo" went from somewhere close to 0.3 to 0.1 (on a scale of 1, where 1 would be 100% of the energy bounced back and 0 would be where 100% of the energy is absorbed) as the ground was disked and plowed for next year's crop.

Suddenly, playing outdoors, running through the fields with the sisters and the dog didn't seem quite as appealing in the 100 degree heat and a 25 mph wind blowing across miles and miles of plowed field, and activities soon turned to being in by the air conditioning as much as possible. As a child, this "air conditioning" meant a big "swamp cooler" hanging in the west window of the house with the cooling coming from the evaporation of water flowing through a mesh of Aspen wood. To this day, I can close my eyes and still remember the smell that created every year when new pads were put in, and although it wasn't as cool as "city-fied" air conditioning, it sure beat the heat in the back yard!

As I grew older my responsibilities increased and I couldn't sit around in the house by the air conditioning all day. I had field work to do on a cab-less tractor or putting up hay, and these were the times that I really came to look forward to the weather of fall, whether it brought school or not!

And sure enough, before I knew it, it was mid-August, time to go school shopping and get ready for another year (that always started within a few days of my August 25th birthday). That long list of everything that I was going to get done in the fun of the long summer days was left incomplete as the dog-days of July and August ate away at all the energy I might have had to get them done way back when I made the list in the cool, beautiful, green days of late May and early June.

The Illinois Cousins' Impression of Kansas in Summer

As a child born to a big family, it was a regular rite of summer to have cousins come and stay at our house for a day or more during their summer trips. We did the same, never paying to stay in a hotel on any of our vacations, but having cousins' houses strategically plotted along our vacation route where we would spend an afternoon or a night on our way to the next cousins' house.

My mom's side of the family, the McFaddens, came from Illinois a generation or so back, so we still had lots of cousins there, and some of them made their living growing corn. Evidently, after the corn is tasseled in July and before they can harvest it in September, they had time to come and visit their Kansas cousins sometime in early August.

While Kansas is a beautiful place to live and visit, if I were to bring someone here to show them the best the state has to offer, I probably wouldn't bring them here in early August. The landscape has more of a brown tint to it than any other color and the ceaseless Kansas wind feels akin to a blast-furnace blowing in your face. This was not lost on the Illinois cousins and they would ask, as nicely as they could, "Why would you live in such a God-forsaken country?" We would assure them that it was not always this way and most of the year was quite enjoyable, but the skepticism in their eyes gave away the fact that they didn't believe us even a little bit. Undoubtedly, this was the time of the year when old Zebulon Pike came through Kansas and wrote back to the president in 1823 to tell him that it was nothing but a "Great American Desert" and that we should consider importing camels to get across it.

One year, after one of our Illinois, corn-farming cousins retired and their children had moved away from home, they figured out they could travel whenever they wanted and made their trek to the Kansas cousins in May. It was one of those years where all of the rains were just right and everything was as green and lush as a postcard from Ireland. The fields of wheat, not yet headed out and still looking like tall grass stretched on for miles and miles over gentle, rolling Kansas hills. The pastures were knee-deep in long-stem bluegrass as they drove through the Flint Hills on their way to see us. The sky between the frequent rain clouds was a deep color of blue and the air was alive with the smell of late spring. I am guessing that this was the time of the year when the Great Plains and Kansas earned their nickname "The Garden of the World" around 1880.

All I remember from that particular visit was how they kept saying over and over what a paradise Kansas was and they finally understood why we would live here and couldn't understand why we would ever want to leave.

Droughts in Kansas

Kansas has gone through drought cycles about every 25-30 years for as long as records have been kept. Before Kansas was even a state, in 1860, a drought hit that was so bad that it caused 30,000 settlers to uproot and leave. Another drought hit in the 1880s shortly after Kansas had earned the nickname, "The Garden of the World" thanks to a very wet year in 1880. Not much is written or said about any drought around the turn of the 20th century, but then came the 1930s.

My dad was a "dust bowl baby" born on a farm west of Dodge City, Kansas in early 1936. His mother had to hang a wet towel over his crib to keep him from getting dust pneumonia and something about the dust caused his ears to not develop correctly. He had ear doctors look in his ears later in life and they said, "you were born in the 1930s, weren't you?" He told many stories that his parents had told him of that terribly hard time to be a farmer in western Kansas and in this part of the world, that drought shaped an entire generation.

The drought my father spoke of more was the drought of the 1950s. In many ways it was meteorologically as bad or worse than the drought of the 30s, but land management practices had improved so much that there weren't the huge dust storms that grabbed all of the headlines in the 30s. The story he told that made the biggest impression on me was of a horse they had that was put out on 30 acres of wheat pasture just to try and keep him alive since the wheat was not going to make any grain. The horse died, and when they did an autopsy to find out why, its stomach was full of rocks from having to go down so close to the ground just to find enough green wheat.

The next drought, and the first one I remember was the drought of the 80s. I wasn't paying a lot of attention to the weather yet at that age, but I knew enough to know that the weather was causing crops to fail and I remember the price of corn setting its all-time record price up to that point.

It is the drought of the "twenty-teens" that I will be telling my grandchildren stories of. The summers of 2011 and 2012 were so terribly hot and dry that record amounts of the country's farmland were declared disaster areas. Grain prices again reached all-time record highs, cattle had to be sold off because the pastures couldn't support them and there wasn't enough hay to feed them. Wildfires burned large portions of both forest and grassland from Kansas and Oklahoma to Colorado.

If history is any indication, the next big drought will hit Kansas sometime in the 2030s or 40s. There will be some good years before that hits, so we should enjoy making the most of it.

FALL

The Cadence of Talk in the Country

As I said earlier in this book, my father and the neighbors that surrounded us were master storytellers and it was always a joy to listen to them tell their stories and laugh. Not much different than two neighbors in the city talking over their back fence, I suppose, but there are two kinds of "speech" from my childhood that I miss and associate intricately with the farm.

The first is the sound of an auctioneer. I know that they still do have auctions in the city, but growing up on a farm, auctions were just a way of life; something you went to several times a year either when a farmer was selling out his equipment or to the livestock barn to sell or buy. Before the farm crisis of the '70s, most towns had their own sale barns with a weekly sale. Since most farms had everything from hogs to goats to cattle to horses, there needed to be a close market for those things to be bought, sold and traded. I remember going to the Kingman Sale Barn with my dad on many occasions, and after it closed, going to the Hutchinson or Pratt sales with my uncle. It is a different kind of auction where they run the animals being sold through the ring, the auctioneer makes a few announcements about the lot, then the bidding begins. For a child, this was a terrifying time. The farmers and ranchers that were bidding would barely nod or blink and somehow the callers on the floor would know that they were bidding. Inevitably, my nose would itch at that exact moment, but I didn't dare scratch it for fear of buying a lot of bred heifers and having no way to pay for them or anywhere to keep them. My father would put the fear in us of acting up at a sale by telling us the story of a time that a man's son ran up a bid on a horse so high that the auctioneer finally had to stop the sale and say, "Now, Mr. Brown! Do you *really* intend to buy this horse because your son has the winning bid at $500 at the present time" after which the red-faced father and sheepish son had to exit the sale barn to the laughter of their friends and neighbors and the bidding had to start all over.

Another rhythmic sound that is inseparable from the farm life in Kansas is the sound of the market reports on the radio. I have heard tell that in England, on BBC radio, they give a nightly shipping weather forecast for the north sea. At one time, they decided to take that forecast off of the air since it applied to so few people, but they were overwhelmed with requests to put it back on. People would listen to the rhythm of the forecast for the different areas of the sea and it somehow comforted them. Perhaps it took them back to their ancestral roots as sailors or perhaps it was just a repetitive sound that helped lull them to sleep. The market reports on the radio are like that to me. I have very little interest in the miniscule day-to-day movements of the grain or livestock markets, but something in me just can't pass by a report on the radio without stopping to listen for a while, "Kansas City July wheat up one and three-quarters to eight fifty-nine and one quarter. September beans down 5 to twelve seventy-five and a half. On The Mercantile, June live cattle down ninety-three to one-oh-five seventy-five..." It is like a language all its own but it is a language that runs in my veins and it is my "ship report" that touches a part of my soul long forgotten.

A Time to Sow

The farm I grew up on was a "traditional" farm with all of the animals and diversified crops up until the farm crisis of the 1970s, so I don't remember our farm that way very much. What I remember more is after dad had to get a job in the metal factory in town. Working 40 hours a week there, he had to become a "specialty" farmer, get rid of the livestock and multiple crops, and just pretty much grow the crop that Kansas is known for--wheat.

Wheat is a relatively easy crop to grow in the semi-arid climate of Kansas and is not greatly labor intensive. This worked well with dad's schedule as he could take his "vacation" at harvest and planting time and then just work around his "town job" the rest of the time.

Hard, red, winter wheat is a unique crop in that it is "backwards" from what most people think of when they think of traditional planting and sowing. Most people think of planting in the spring or summer, then harvesting in the fall like you do with corn, soybeans, milo and about everything in your flower or vegetable garden. Winter wheat, however, is in the grass family, so it is planted in the fall to get established, goes semi-dormant over the winter and then is harvested in the late spring/early summer.

An entire culture is built around the fall harvest with harvest moons, songs, poems and holiday displays, but to a kid growing up on a wheat farm in Kansas, fall meant not taking the crop from the ground, but putting the seed in with hopes of a bountiful harvest nine months later, and this is one of my favorite times of the year.

After a long, hot summer of working the fields to keep the weeds down as they sat fallow, the cool weather of late September was so refreshing. Riding the cab-less tractor in July and August just drained the energy out of you, while the coolness of the fall was invigorating. There was also something about the pace of life when drilling wheat that was different than the pace of harvest or plowing. Yes, it was imperative that you got the seed in the ground, but it wasn't like you were racing against time or approaching storms.

Aesthetically, the drilling was a pleasure to the eye, too. All of the big clods of dirt that had been turned over by the plow had been smoothed out throughout the summer, and when the drill was done moving over the land, it looked like a soft, smooth, brown quilt with the lines from the drill stitching the fabric of the soil to the earth.

There is also something about the smell of dirt that smells different when smelled trough cool, fall air versus scorching summer temperatures. It is one of the most comforting, earthy smells and to this day, if I drive by a field being worked in the fall, the smell instantly takes me to drilling wheat on the farm.

My dad used certain signs to know when the soil temperature was cool enough to plant. Evidently, if the soil is too warm from the summer, the seed won't germinate as well. The sign my dad looked for was the "spider webs." I don't know if they actually came from a spider or some sort of a silkworm or some other creature altogether, but as the soil reaches a certain temperature in late September, you can look across a dirt field toward the sun and see millions of silk threads stretching across the ground. Dad would see that and say, "Now it is time to plant." Whatever creature created these webs evidently worked quickly because after drilling or working a field, the next day, the ground would be covered with those webs once again.

Since you wanted to make sure your drill was working properly and that you didn't miss any rows, the drilling days were shorter than the harvest days as dad never wanted to drill without sunlight. There is nothing like the feeling of turning off the tractor at the end of the day with the coolness of a September evening, full of hope that the seeds you just put in the ground would grow to a great harvest next year.

October

My favorite month of the year. Not too hot, not too cold. Some crisp, cool evenings when a jacket feels good. Warm soup. Poetry. Family gatherings. Friday night football games. Marching band competitions. Changing leaves. Outdoor projects. The first frost. Trips to the farm. A hearty appetite. Homesickness. My father's death. The farm. Hedge apples. Warm socks. Blankets. Lighting the furnace. The smell of the dust burning off in the vents. Draining the hoses. Checking the antifreeze. Bales of hay. Pumpkin patches. Candy. A hay-rack ride. Thickening hair on the dog. Deer in the early evening twilight. Geese flying over. Sandhill cranes crying out from on high in the shape of a "V". Ghost stories. A quiet house. Long evenings. The World Series. Bright, blue skies. Foggy mornings. The Earth sighs. Leaves fall. Busy squirrels. Cutting and splitting firewood. Drilling wheat. Slowing down the pace of life. Kids in costumes. Trick-or-treating. Getting to dress up as something you always wished you could be. A Jack-o-lantern with a real candle in it. Coyotes howling. A full moon. Hot apple cider. Warm cookies. Chili. Finishing up outdoor projects. An early gale. The longest month seems too short.

"When the Frost is on the Punkin"

It has become almost a cliché for fall, but that line is from a poem that is very precious to my family and me. Written over 100 years ago by Indiana "Hoosier" poet James Whitcomb Riley and contained within a small, orange book called, "Riley Farm-Rhymes," it has become part of a tradition in my family that I wouldn't trade for anything.

My parents grew up in a time before television where the main forms of entertainment were radio, reading and story-telling. Their parents would read to them from giant, well-worn books of poetry and occasionally one of those poems would speak very deeply to one of them. As we grew up in the 70s, this form of entertainment had been pretty much replaced by the television, but there were certain poems or lines from poems that my parents would quote at an appropriate time.

Suppertime at our house with all nine of us around the big table our father had built was a nightly, family event. Television and radio were not allowed and conversation was encouraged, often long after the last bite had been eaten. We would sometimes take turns "going around the table" telling about our day, and everyone else was expected to respectfully listen to the other. Occasionally, when it was Mom or Dad's turn, their conversation would turn to a line or stanza of poetry that they remembered from their youth, and we would have to wait patiently while they went to Aunt Mag's bookcase to get a poetry book and find the poem so that they could read it to us in its entirety.

That is how the family tradition of reading Riley's poem came about. It would be a cold, crisp Autumn day and my dad would come back after doing chores and say, "Tonight would be a good night for that poem." After supper, he would go to the bookcase, take out the small, orange book and read the entire poem of "When the Frost is on the Punkin" with its images of fall and gathering crops and heaps of apples and if the Angels came around and wanted "boarden'," that this was the time of year he would want to give it to them.

As we kids grew older, we came to expect that poem every year on a crisp Autumn day, and when we started going off to college, we would call home and say, "Don't read the poem until I am home this weekend," and a tradition was born.

Every year, usually sometime around late October, all of the kids would come home to a meal of soup or some other warm, fall, comfort food and after supper, we would sit around as dad read the poem. Occasionally, others would read other poems with some of the favorites being "October's Bright Blue Weather" by Helen Hunt Jackson, "America for Me" by Henry Van Dyke, "Little Orphant Annie" by James Whitcomb Riley, and if we were in a sad or sentimental mood, "The House with Nobody in It" by Joyce Kilmer, "Rock me to Sleep" by Elizabeth Akers Allen or "Jenny Kissed Me" by James Henry Leigh Hunt. Sometimes, if we were in a silly mood, we would pull out some like, "Methuselah's Diet" by Anonymous. No matter whether it was just one poem or a whole evening of poetry, it was a night we all looked forward to.

Now that we are all grown and have kids (and grandkids) of our own, we perhaps look forward to the "Frost Festival" more than ever. For many of us, it is our favorite "holiday" surpassing Thanksgiving or even Christmas. The date is set a year in advance, we all bring our favorite soups or desserts and have a wiener roast for the kids (if the Kansas wind isn't too strong).

There were a handful of years where we didn't have our "Frost Festival." When my father got sick with prostate cancer, he decided he was not up to it and it was time for us to take the tradition to our own families, so we got together as individuals and read the poem and it looked like the tradition was going to be passed on. The next year, however, my brother died, and we all decided we *needed* to be back together again, but it was too painful to read any of the poems (he was the "poet" in the family), so we just comforted each other with food and conversation. The year after that, dad died in October, right around the time the festival would have happened. We were all together for the funeral and although we didn't read the poem, we had it printed on the back of his funeral program. Those that knew of the tradition in our family understood the poignancy of this and I liked to picture my dad meeting those choirs of Angels with these words:

"I don't know how to tell it--but ef sich a thing could be
As the Angels wantin' boardin', and they'd call around on *me*--
I'd want to 'commodate 'em--all the whole-indurin' flock--
When the frost is on the punkin and the fodder's in the shock!"

www.ingramcontent.com/pod-product-compliance
Lightning Source LLC
Chambersburg PA
CBHW071802200526
45167CB00017B/1090